HOW TO BUILD MAX-PERFORMANCE FORD FE Engines

Barry Rabotnick

CarTech®

CarTech®
CarTech®, Inc.
39966 Grand Avenue
North Branch, MN 55056
Phone: 651-277-1200 or 800-551-4754
Fax: 651-277-1203
www.cartechbooks.com

Edit by Paul Johnson
Layout by Monica Seiberlich

ISBN 978-1-934709-15-3
Item No. SA183

Library of Congress Cataloging-in-Publication Data

Rabotnick, Barry.
How to build max-performance Ford FE engines / by Barry Rabotnick.
p. cm.
ISBN 978-1-934709-15-3
1. Ford automobile—Motors—Design and construction. I. Title.

TL215.F7R33 2010
629.2504—dc22

2009052232

Printed in China
10 9 8 7 6 5 4 3

Cover:
Barry Rabotnick entered this 427 high-riser engine in the 2009 Engine Masters Challenge. It features a Genesis block, forged-steel crank, set of Carillo H-beam rods, and CNC-ported Blue Thunder heads with 11.5:1 compression. The engine cranks out 670 hp on 91-octane pump gas. (Courtesy of Robert McGaffin)

Title Page:
This FE engine is adorned with the factory 2x4 carb and intake package, which has tremendous cosmetic appeal and performs quite well, even on mildly modified engines.

Back Cover Photos

Top Left:
The front of the renowned FE 428 block is shown, and from this angle it looks identical to other FE engines. The 428 fitted to Torinos, Mustangs, and Galaxies became a legend on the streets.

Top Right:
Installing a Scat crankshaft is a popular aftermarket upgrade. The Scat 4.250-stroke FE crank is a proven durable, ductile iron piece, which is perfectly suitable for any street use as well as moderate race applications. These can handle about 700 hp with no issues.

Middle Left:
Here's a close-up view of the C8OE-N Cobra Jet head's combustion chamber.

Middle Right:
This is a typical roller rocker arrangement for an FE engine, and it shows proper alignment as well as valve-tip contact.

Bottom Left:
This 2x4 carburetor setup sits on a Tunnel Wedge intake. This classic factory-style combination is a viable option for many FE engines today. While this intake may be a better option for a race engine, certain engine combinations provide acceptable street performance characteristics.

Bottom Right:
In this photo, the camshaft is carefully guided into the block. You can use both hands to smoothly slide it into place. You need to be careful to avoid banging it on the block or any internal components.

CONTENTS

ACKNOWLEDGMENTS

Writing a book is a solitary endeavor, but researching and gathering photos for this book required the help of a number of individuals. In particular, I thank the entire Ford FE Forum team with a special thanks to Jay Brown, Dave Schouweiler, and John Vermeersch. They provided valuable assistance in the creation of this book. My staff at Survival Motorsports helped me as I deviated from my traditional work duties, so I could complete the book. My guys spent a lot of time breaking up their normal work routine to accommodate the necessary engine build process and photo shoots. And they put up with me when I was writing the manuscript. These helpful and conscientious colleagues include Tom Gunther, Valerie Simon, Marc Wiener, and Tim Young. I thank Acquisitions Editor Paul Johnson. His calls and emails helped shape the manuscript. His patience and understanding was very appreciated. And of course, I couldn't have done this book without the support of my home team. Susan, Autumn, Summer, and Lily—thanks for all your love and support during the process. Without all of you, it wouldn't have been possible.

INTRODUCTION

The FE engine was one of the most prolific Ford V-8s to ever grace an engine bay. Millions of FE engines found their way into Galaxies, Fairlanes, Mustangs, Torinos, Thunderbirds, trucks, and other vehicles. A lack of high-performance parts relegated the engine to a "has-been" status that endured until Edelbrock released high-performance aluminum heads in the mid 1990s. Since then, the FE engine has enjoyed a renaissance and today, every variety of FE engine part is available in the aftermarket. You can build an entire max-performance FE engine without using a stock part. Wow, how times have changed!

With a massive selection of aftermarket and stock offerings, there's a huge range of parts to choose from. This book will help you sort it all out, so you can build an FE engine with compatible parts to meet a particular performance target—blocks, heads, valvetrain, cams, exhaust, ignition, induction, and everything in between.

I also want to mention what this book will not do. With an engine that has been used for so many purposes, and with such a storied legacy, the challenge lies not in getting enough information, but in assigning some limits; therefore, I cannot cover everything.

This is not a history book detailing each original application or a fabled race car. There are other far-more-qualified folks to write those stories. Many of the guys who raced during those glory days are still with us today. My historical information is brief and limited to the first chapter, but it is needed to provide context.

This book discusses all relevant components and procedures for building strong, reliable, high-performance FE engines. But it isn't a comprehensive source for production engine and component data.

This volume is focused on high-performance street applications, and it is not a builder's guide to FE race engines. These are engines built for more power in a street-driven car or truck, as well as racers seeking a competitive edge in a bracket car or road-course Cobra. I give a nod of respect to the original FE race programs, use the restoration information as a backdrop where appropriate, and look to the NHRA-class racers and builders as an inspiration. This is a guidebook for the "man in the middle"—the hot rodder.

Today's sportsman racers are a tight-knit family of skilled professionals. They have perfected the art of getting the absolute most out of a given combination, through years of continued development. Within the Super Stock or Stock Eliminator ranks, an FE engine is still considered a competitive piece of equipment thanks to their dedication.

CHAPTER 1

A Brief History of FE Ford Engines

This 520-ci FE with 12.5:1 compression cranks out 770 hp and 653 ft-lbs of torque at 6,900 rpm. It features polished high-riser heads and a Tunnel Wedge manifold with dual-quad carbs. This is about as nice as they come.

You can easily build a 500-hp stroker FE engine these days. Case in point, this 475-hp 445-ci is based on the 390 FE and delivers a reliable 475 hp under 6,000 rpm. No grinding or clearancing is required to assemble the Survival Motorsports stroker kits into an FE block.

The FE Ford engine was released into production in 1958. The earliest applications included use in the short-lived Edsel program. The FE was not a replacement for the Y-block; it was a larger companion to an engine family sharing some design features. In 1958, the Y-block was still considered a current design at only four years old.

Starting out at 332 ci, the FE quickly grew in displacement through its first five years of production, with 352-, 390-, and 406-ci variants followed by the now famous 427 in 1963. By 1966, the renowned 428 and the short-lived 410 had been released, and these completed the lineup of FE passenger-car engines. And as a result, a lot of high-performance engine history was written in a very short time. The 352 and 410 were dropped after 1966, and the 390 and 428 continued as the only FE engines in passenger-car production from 1968 through 1970.

The FE had been dropped from passenger-car use by 1971, but the 360 and 390 versions remained extremely popular in pickup trucks through the 1976 model year. Some commercial applications, most notably U-Haul trucks, had FE power through the 1978 model year. Throughout the 20-year production run, the FE had seen use as a marine, commercial, and industrial engine as well.

The high-performance factory engines were the ones that claimed all the glory, but the vast majority of engines were for more mundane applications. The most popular original FE vehicles were full-size family cars and pickup trucks, and these vehicles contain the engine blocks that are used for many high-performance engine builds today.

The beginnings of the FE performance program took place when Ford split the car lines during the late 1950s, going from one basic platform to many as the market developed. The emergence of the bigger cars coincided with a gain in the popularity of racing. The NHRA U.S. Nationals were held at Detroit Dragway in 1959 and 1960, and auto executives were exposed to the rising popularity of the sport. At the same time, NASCAR began the transformation that would take it from being a local-circuit group to a national sport. Television was about to change the way cars were marketed, and motorsports was one of the beneficiaries.

Ford responded to the market opportunity with high-performance iterations of the 352 and then the 390. In this era, a production-based engine could still be equally successful in drag racing and NASCAR.

The FE performance program started out as upgrades to the passenger-car engines, using strategies that had been employed by hot rodders for several years. Higher compression, multiple carburetors, and dual exhaust were initially enough to get attention. But as the rivalry between the "Big 3" heated up, they quickly evolved into performance-specific engines. The first of these was the 406, blessed with a larger bore than the 390, solid-lifter cams, and optional multiple carbs. Within a couple years the 427, with a still-larger bore, cross-bolted main caps, and better cylinder heads, replaced the 406. The 427 became the lead piece for all of Ford's big-block race development, and remained in that position through the end of direct factory involvement in 1970. When discussing professional racing and FE engines, the 427 is going to be the focus of conversation.

My 427 high-riser entry as campaigned at the 2009 Engine Masters Challenge. The engine features a Genesis block, a forged-steel crank that came out of a truck, and a set of Carillo 6.700-inch H-beam connecting rods. On top of the 427 is a set of CNC-ported Blue Thunder heads with a 11.5:1 compression ratio. This engine made more than 670 hp on 91-octane pump gas in contest trim with a .697-inch lift, solid-roller cam and two 750 Quick Fuel modified carbs.

The 428 was originally released in 1966 as a torque-oriented street engine. But in the late 1960s, somebody at Ford finally realized that the low-production and high-strung 427 was not reaching the masses. Ford had a good race program, but it was getting a bad street "rep" because the more-mundane 390-powered cars could not keep up with the GM or Chrysler big-blocks. The response was to blend the readily available and bigger 428 blocks with higher-performance parts, which included heads, cam, and intake. The 428 Cobra Jet package was available from late 1968 until 1970. It delivered on all points and thus provided a reliable, strong, and still-competitive combination in NHRA class racing.

The 429-engine family was slated to replace the FE, but the factory programs surrounding the new engine were short lived, barely making it two years before performance development stopped. Eventually, the potential for the "385" family engine was realized, but that is another book.

The Famous Cars

Ford's initial platform for FE performance and racing was the full-size cars, the most popular being the higher-end Galaxie. Many FE engines were installed in full-size cars, most of them 352s and 390s. But the racers got the 427 cars.

The 427-powered Galaxie was a competitive package, but the Chrysler cadre had a distinct weight advantage with its smaller cars. The first response was to develop a lightweight factory drag-race version of the 427-powered Galaxie. It included a high-riser version of the 427 engine, along with a variety of weight-reduction strategies, including changes to sheetmetal, interior parts, and even the frame. Always rare, and quite valuable today, the lightweights were only the opening act.

The next step was a factory-authorized, dedicated drag-race car: the Fairlane Thunderbolt. Dearborn

This 428 FE engine is dressed out for street use in a 1966 7-liter Galaxie. The 428 featured an externally balanced, cast-nodular-iron crankshaft. Because of longer stroke, hydraulic lifters, and reduced compression ratio, the 428 was much more streetable than the 427. Affordable 428 FE engine blocks are a rarity these days. If you see one at a swap meet or on an online auction site, you need to check inside the water jackets as well as the casting numbers to confirm that it is indeed a 428. Some unscrupulous sellers have overbored 390s and tried to pass them off as 428s.

As the factory horsepower wars heated up, even the lightweight 427 Galaxies were deemed too heavy to run with the mid-sized Dodges and Plymouths with their 426 Hemi power. Ford partnered up with Dearborn Steel Tubing to produce the 1964 Fairlane Thunderbolt to regain the competitive edge. The 427's high-riser manifold necessitated the use of the bubble hood. While it looked like a road-going Fairlane, the Thunderbolt was a genuine race car. Rear window cranks, windshield wiper, carpeting, radio, heater, sound deadener, and body insulation were all deleted to save weight. Most cars posted quarter-mile times between 11.6 and 12.0 seconds. Still competitive in Super Stock today, Ray Paquet and Paul Adams have run T-Bolts into the 8-second quarter-mile range.

Steel Tubing, a Ford contractor, assembled the T-Bolts. It took the lighter-weight, mid-size, 1964 Fairlane sedan and installed the high-riser 427 engines into about a hundred of them. This was never intended as a street vehicle, and everything was modified to enhance the cars' chances at the drag strip. This included major front-end work to accommodate the large engine, lightweight seats, thin glass, aluminum and fiberglass components, and a race-only rear suspension. The Thunderbolt became a Ford racing icon, and the combination remains near the top of NHRA Super Stock racing 45 years later.

Ford did not install the 427 in production Fairlanes until 1966. The production 427 Fairlanes from 1966 and 1967 are both very rare and very competitive cars, with a solid racing history. But like the lightweight Galaxie that preceded it, it never received the adulation reserved for the Thunderbolt.

Something about the almost absurd combination of small car and huge engine makes anything else seem normal in comparison. The ultimate expression of small car/huge engine is also FE powered—the 427 Cobra. The Cobra started out as the well-documented combination of a British sports car and a Ford small-block V-8 for road racing. The roadster competed with well-funded efforts from domestic and foreign racers, and the 427 FE, a readily available race engine, satisfied the need for more power. What had already been an attractive sports car morphed into a beauty born of necessity, with broadened and flared fenders for larger tires, side exhausts, and a scooped hood. Brutal in both potential and execution, another automotive icon was born. Today, there are many, many more inspired iterations of the car available than were ever originally made.

NASCAR racing was the primary development and test bed for Ford's FE race program throughout the 1960s. The 427 was upgraded and altered every year as needed to remain competitive. But while NASCAR served as the engine technology source, the cars themselves were not inspiration for many production performance offerings. Muscle-car enthusiasts and street rodders looked to NASCAR for entertainment, but to the drags for inspiration. So while we use parts that were designed for the high banks, we don't emulate the cars themselves very often. Street cars have the big tires on the rear, scoops on the hood, but no numbers on the doors—a tradition that still holds true today.

Throughout the late 1960s, professional drag-race programs evolved and the cars got further from a production basis. The hard-core drag racers moved into AF/X cars, with radical modifications to wheelbases and engines. These in turn morphed into "Funny Cars," which used tube chassis and nitromethane fuel. The SOHC FE or "Cammer" engine remained a common powerplant in these exotic race machines, but it was far removed from the engine you'd get in your car from the local dealer. These cars and engines are certainly worthy of discussion, impressive by any measure, but beyond the scope for this book.

The most famous of the FE-powered cars was never really sold to the public. Ford made a very public and concerted effort to win the 24 Hours of LeMans race in the middle 1960s. Ford put enormous resources behind the effort because the company wanted to break the stranglehold that Ferrari had at LeMans and establish itself on the international racing stage. Enzo Ferrari's scarlet cars had won the race from 1960 to 1965, but that was about to end.

To start with, Ford used the small-block V-8s to power the GT40 sports racing cars. In subsequent years, the need for more power became apparent. In a situation similar to that of the Cobra, Ford opted for the well-developed 427 FE as a power upgrade to the GT racing program. And the engine delivered; Ford GT40 cars finished 1-2-3 in 1966. But perhaps the most memorable win came the following year, as legendary American drivers A. J. Foyt and Dan Gurney won the 24 Hours of LeMans in an American sports race car, the GT40. Most notably, the FE 427 powered Ford GT40s to four consecutive LeMans wins from 1966 to 1969, an epic achievement for Ford and the FE engine.

So here is the FE engine legacy: It was the engine that was in the most famed Ford racing vehicles of the time in each form of motorsports—NASCAR, the Cobra, the GT40, and the Thunderbolt. This should be the backdrop for comparable fame and popularity on the streets of America, but it never happened. What went wrong?

Mustangs, Galaxies, Fairlanes and Trucks

As a dedicated Ford fan and a Detroit-area FE racer since the 1970s, it hurts to say this but it needs to be said. What went wrong is that Ford put everything into the low-volume, high-dollar racing efforts and comparatively very little resources went toward the everyday cars that made up the greatest volume of production.

The FE was factory installed or available in numerous car and truck platforms. The full-size Galaxie (and sister models) was the recipient of most FE production, from the early 1960s right to the end. Most popular among enthusiasts are the 1963–1967 models.

Ford intermediate cars, the Fairlane, Torino, and Mercury variants from 1966 through 1969 had the FE as a regular production option. Most were 390 powered. A very few 1966–1967 models had a 427, and the 428 CJ was available in 1969.

Mustangs and Cougars were often FE equipped from 1967 through 1970. The 1967 and 1968 big-block models were nearly all 390 equipped. In 1969, there were a few 390s, but the 428 CJ was the engine of choice. The hydraulic-lifter version of the 427 was installed in a few Cougars in 1968, but no 427 Mustang has ever been documented, despite 30 years of rumors.

Ford pickup trucks carried the FE as an available option through 1976. There are probably more FE engines in pickups than in any of the cars. The FE can be installed into any of the cars or trucks where it was an option. Any deserving small-block or 6-cylinder-powered candidate can be converted to FE power using factory replacement components.

When new, a 390-powered Galaxie of 1964 or earlier was a competitive car on the streets and local tracks. But by the 1970s it was common knowledge that the average 396-powered Chevelle was significantly faster than any 390 car. A 428 Mustang could hold its own, but the majority of FE owners simply lost enthusiasm because they were outgunned every Friday night. They moved on to other cars or other hobbies, and the FE-powered cars were left to sit or be used as basic transportation. Interest from the aftermarket never really took off, so the supply of new parts was not there, and the old factory parts were getting used up and worn out.

By the 1980s the FE engine was considered obsolete by all but a few die-hard enthusiasts and racers. No mainstream magazine coverage, no new aftermarket parts, and no real development existed outside of the private effort of a few NHRA Super Stock racers. The engine design that had won Daytona, LeMans, and the Winternationals was considered to be in the same league as the Buick Nailhead, the Chevy 409, the Olds Rocket, and the Y-block.

The FE Reawakens

But there was a difference: the cars. The Cobra was still worshipped, the Thunderbolt was still an icon, and the legacy from those early NASCAR and drag-racing wins still hung on. Stock and Super Stock racers running FE power continued to win with no factory support. As people started to repair, reproduce, and emulate those cars, the demand for FE parts began to build.

Specialty suppliers, such as Dove, carried the FE torch through the slow years, catering to the dedicated racers and restorers. Demand started to build in the mid 1990s when Edelbrock released a replacement FE aluminum cylinder head. Equally important, there were a lot of candidate engines available from the huge truck population, and there were also a lot of candidate cars to choose from.

In 2004, Scat released a cast-stroker crank for the FE, and Genesis concurrently released the cast-iron reproduction 427 blocks. I built the very first big-inch FE engine that used both parts, topping the 505-ci package with an electronic fuel injection (EFI) system. The engine was profiled in *Hot Rod* magazine's July 2004 issue as the "676 Horsepower Dinosaur."

I entered a similar 505-ci FE in the Jegs Engine Masters Challenge the following year, using the new Blue Thunder cylinder heads. Most of the competitors thought it was pretty cool to see one of those ol' FE motors in the contest, and at first viewed it as a curiosity. It became apparent that this was not a nostalgia piece when it made 752 hp on the dyno with pump gas. Essentially, it was a modern engine with FE architecture. I finished eighth overall out of 50 entrants, and got another magazine article as a result.

Jay Brown from Minnesota entered his FE-powered 1969 Mach 1 into Hot Rod's Drag Week competition in 2005. This is a grueling event covering more than 1,000 miles and

The 2006 Engine Masters Challenge entry was a 427 with a belt drive, flat-tappet cam, and a single Quick Fuel 1050 carb. It made well over 650 hp on 91-octane unleaded. This package consisted of a highly modified 391 truck crank from Performance Crankshaft, a set of Scat 6.49 H-beam rods, and custom 10.5:1-compression Diamond dish pistons. Heads were custom CNC-ported Blue Thunder castings with an extensively modified Dove intake.

A Novi 2000 blower on a stroked 428 FE delivers more than 650 hp with roughly 10 pounds of boost. The system runs reliably on 93 octane without risk of detonation. The engine started out with the addition of cross-bolted mains to a factory 428 block, along with custom 9:1 pistons and a 4.250-inch-stroke crankshaft. It is topped off with some mildly ported Edelbrock heads and an owner-fabricated blower system. The fuel-injection conversion uses a modified intake with bungs for the injectors, which are welded into place. In addition, a F.A.S.T. management system handles the electronic aspects of fuel induction.

The legendary 427 single overhead cam Ford, called the SOHC or "Cammer," was never installed in a production car. The 427 SOHC engine powered many Ford drag racers to victory throughout the 1960s and 1970s. This is the most exotic FE engine that Ford built in any sort of numbers. It features a forged-steel crank, forged-steel connecting rods with capscrew fasteners, hemispherical domed pistons, and many other trick parts. The heads flowed an incredible amount of air, featured huge intake runners, and had D-shaped exhaust ports.

The FE designation stands for Ford Edsel. This series of engine was first released in 1958 and continued in large-volume production until 1976. This 1967 427 medium-riser engine has dual-quad carbs. It features original exhaust manifolds and trim. Ford offered two iterations of the 427 block–side oiler and top oiler. In 1965, Ford released the side-oiler block, which routed oil to the main bearings and then to the cam and valvetrain. The top-oiler version sent oil to the cam and valvetrain first and then down to the main bearings.

five drag strips over a five-day period. The best overall-average ET wins, and the Mach took home the class win. He just repeated the feat in an SOHC-powered 1964 Galaxie.

Subsequent FE race wins, engine builds, and project cars have received an increasing amount of media coverage from writers looking for "something different." With a full array of parts now available, it is possible to build a complete 427 FE from scratch using no original pieces. You can build a 445-ci 390-based FE stroker that'll get you 500 honest horsepower without breaking the budget. In a few short years, the FE engine has gone from near extinction to mainstream again. This is without question the best time in the 40-year history of the FE to be building one for the street.

This fuel-injected 445-ci stroker makes more than 500 hp and is equipped with polished heads and intake for show-car use. Starting off with a basic 390 block, we added a Survival 4.250-stroke kit consisting of a Scat crank and rods, Probe 10.8:1-compression flat-top pistons, and 6.700-inch-long I-beam Scat rods. Heads were lightly modified Edelbrock pieces, and the cam is a fairly aggressive .668-inch-lift solid roller for that "sound." The EFI system starts out with an Edelbrock Victor intake. Comp Cams F.A.S.T. division supplies the fuel-handling components, which include injectors, throttle body, wiring, and sensors. This combination delivers a race-car sound, serious power, and still maintains good manners due to the EFI system's ability to control start quality, idle speed, and part-throttle behavior.

CHAPTER 2

ENGINE BLOCKS

Here's an average FE block; this is the way you find them when searching junkyards and classified ads. Before sinking a lot of money into rebuilding a block, you should Magnaflux the block to check for cracks and have it sonically tested to determine the thickness of the bores. Once you have selected a structurally sound block, the rebuilding process can begin.

This chapter focuses on the foundation element of an FE Ford engine build—the block itself. The FE engine was in continuous production for roughly 20 years, so there are a lot of engine blocks in cars, garages, and junkyards. As the popularity of the FE engine has re-emerged, it seems that every one of these has magically become a 428 Cobra Jet, a Shelby part, or a "survivor" of some sort, even if they started out powering an F-150. While the focus of this book is performance building and not "numbers matching," a certain amount of detective work is mandated when embarking upon any FE engine project.

FE Block Architecture

All FE engine blocks share many common design features, which serve to separate them from the other Ford V-8 engine families:

- 10-bolt cylinder-head pattern using 1/2-inch fasteners
- 10.17-inch deck height as measured from the crankshaft centerline to the cylinder head mounting surface
- 4.630-inch cylinder bore spacing
- Deep-skirt "Y"-block design where the oil pan rail completely encases the crankshaft and main bearings
- Unique bellhousing pattern that isn't shared with any other Ford engine
- A 2-bolt motor-mount pattern for pre-1965 blocks
- A 4-bolt mount pattern for later blocks, which can be retro-fit into earlier applications

The Blocks: Identification and Application

Ford FE engine blocks used for performance builds are generally selected from one of three groupings: 390, 428, and 427. While other blocks are out there, these three are the foundation for the vast majority of high-performance street and track applications.

The common single-web-style block is found in most 360 and 390 engines. The vast majority of FE engine blocks have the straight two-bolt main bearing caps. However, the 427 engine had cross-bolted main bearing caps for greater strength.

Another view of the block shows the single main web reinforcement in detail. The thick main webs and deep skirt design provide excellent strength and help make the FE a very durable engine. It's interesting to note that the latest GM LS series and Ford modular series engines share this block skirt configuration, as well as having the cross-bolted mains pioneered by the 427 FE.

390 Blocks

The 390-based blocks, which have an original bore diameter of 4.050 inches, are by far the most plentiful. These were in production from the late 1950s through the middle 1970s, with the majority being used in large passenger cars and pickup trucks. The common 360 engine also utilized the 390 block, with no difference in features or markings, as did the 1966 410 Mercury. The blocks from medium-duty trucks are very similar to the 390 block, and are referred to as 361 or 391 engines. The medium-duty truck blocks have a larger distributor shaft hole, requiring a bushing for passenger-car use.

There are several differences in 390 engines from various years and applications. Perhaps the most obvious variance is the use of a double or reinforced main web design on the heavier-duty versions of the block. Most of the "mirror 105" blocks (see the casting information that follows), as well as the 361 and 391 medium-duty truck blocks carry this feature. It can also be found almost at random on other engines. If you have your choice, it's probably the better piece, but the benefits are modest at best.

428 Blocks

The 428 blocks, used in numerous Galaxies, Thunderbirds, and high-performance Mustangs and Fairlanes, are far less common than the 390. These blocks are the basis for the famed 428 Cobra Jet engine and use a 4.130 basic bore dimension. Many 428 engines were also used as industrial and irrigation power plants. The Cobra Jet and industrial blocks usually have the double main webbing.

427 Blocks

The rarest and most desirable of the FE blocks is the revered 427—with either center oiler or side oiler. With a base bore diameter of 4.233, this was the basis for the engines found in cars such as the Thunderbolt or the NASCAR program. Most–but not all–427s have screw-in-type core plugs, a feature not found on any other FE engine. With very few exceptions, 427 blocks have cross-bolted main bearing caps, using 3/8-inch fasteners through the sides of the block to add significant strength to the FE bottom end. The center-oiler design uses the same lubrication strategy as that employed in more common FE engines, while the side-oiler design

The standard single web provides essential support for the main bearings and crankshaft. Also visible is the annular oil grooves in the cam bearing bore. This groove feeds oil to the rocker assembly. The groove can be added to side-oiler blocks to permit use of less expensive and more readily available cam bearings.

has a unique main-feed galley along the side of the block (hence the name). Side oilers can be said to prioritize main and rod oiling, with upper-end lubrication being transferred through grooved camshaft journals (more on this later).

Most 427 blocks were used in high-performance passenger cars and were never produced in large volumes. Many others can be found in marine applications, but rarely in industrial engines. Brass-core plugs usually identify a 427 as a marine engine. Earlier 427 engines were all equipped with solid lifters, and have no provisions for lifter oiling; conversion for hydraulics is possible, but difficult and expensive. Many of the service blocks, the few used in the 1968 Cougar GTE, and the marine blocks usually have hydraulic-lifter oiling provisions. In addition, many marine 427s are cast as side oilers but drilled as center oilers.

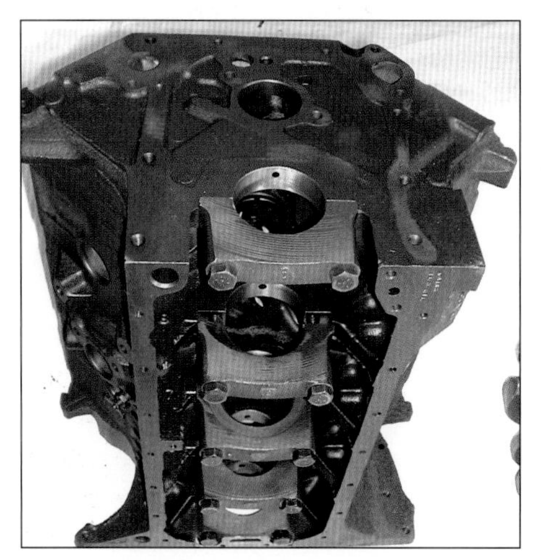

The 428 block bottom is visually no different than a 360 or 390. You cannot go by appearance alone when trying to identify an FE block. The vast majority of blocks produced were used in trucks and started life as 360s or 390s; the 428 is a pretty rare part in comparison.

The casting numbers as shown on this 428 are useful for date code, but do not tell you what the engine actually is. Casting numbers on FE blocks are most often useful for exclusion, not for identification. As an example, a block with a 1964 date code cannot be a 428 because the 428 did not come out until 1966. Some FE blocks had a partial VIN (vehicle identification number) very lightly stamped into them, but even this is not a hard-and-fast rule. An FE hunt is going to require a dial caliper, a flashlight, and some patience.

External Identification: How Can You Tell What You've Got?

The short answer to this most frequently asked question is: You can't, at least not from the outside of an assembled engine in the car. All FE engine-block castings appear nearly identical, with the notable exception of the side-oiler passages and cross bolts above/along the oil pan rail on many of the 427s. It takes a highly trained eye to notice the small cues that define a high-value part from the mundane. And even the best parts scrounger often cannot make a positive identification from markings alone on an assembled engine.

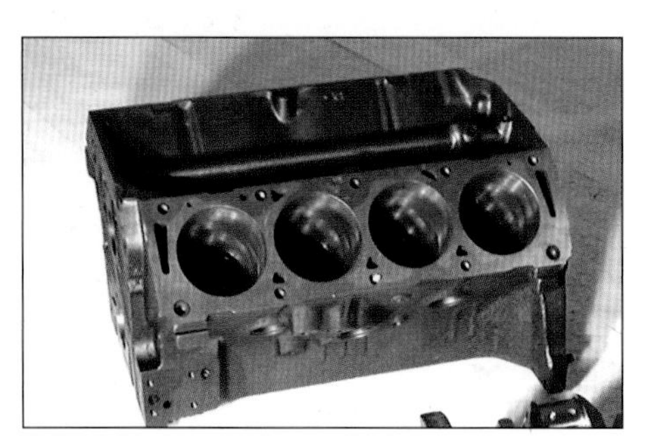

Passenger, service, or industrial 428 blocks often share a common external appearance. Many service blocks have vertical ribs on the exterior, and while these ribs do not add any appreciable strength, they likely make for a nicer pour during the casting process.

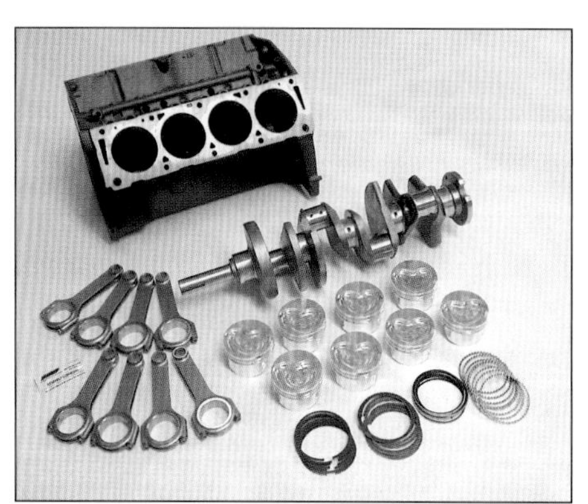

A 428 block and a complete stroker rotating assembly are shown. No grinding or block clearancing is required to install a 4.25 stroke into an average FE. The stroker kits take a normal 390 up to approximately 445 ci, a 428 gets to 462, and a 427 reaches 482 cubes. At Survival we generally use a Scat cast crankshaft in most of the builds, which has proven to handle more than 750 hp with no issues. Diamond or Probe pistons, Scat 6.700-inch-long rods, race bearings, and plasma-moly rings that are file fitted are included in the rotating package. The 390-based engines almost always exceed 500 ft-lbs of torque with power between 450 and 500 with pump-gas-friendly compression of under 10:1. The bigger 427-based combinations have gone over 700 hp.

The "427" cast into the lifter valley it looks cool, but it means little in terms of block identification.

A 427 marine block showing a side-oiler casting has been machined flat for motor-mount clearance. The block is actually a hydraulic-lifter center-oiler design. Conversion to a side oiler by drilling is often impossible due to the machining.

The factory 427 main web is shown in detail. You need to carefully make note of the engraved markings on the spacers and their location because each spacer belongs in a particular location. They should have about a .001-inch press fit and can be fabricated if missing. However, line honing the mains will be required afterward.

Casting Marks

A good place to start your identification search, and a way to eliminate certain possibilities, is with the casting marks on the block. FE engine blocks usually have several casting numbers, both formal and sand scratches, on various areas of the block. Some of these marks are good for identification, but unfortunately many other markings were used almost at random and have little if any meaning for actual identification. I'll cover the most common ones below, but remember that nothing on an FE can be taken for granted. There are actual non-cross-bolted 427 industrial engines as well as paper-thin 390s sold as standard-bore 428s online. Take *nothing* for granted.

A typical center-oiler engine showing the angled feeds for the lifter galleys, which can be restricted if desired. We used to block them off on solid-lifter engines, but do not do so anymore. Restricting the oil feeds raises oil pressure and serves to keep the lubrication volume focused on the crank's rod and main bearings.

The factory 427 side-oiler block is incredibly strong and able to support substantial modification. The side oiler found its way into some of the greatest race cars and super cars of the day, including the GT 40 and Shelby Cobra 427. Here, you see the webbing from another angle, and it's similar to the double-web FE blocks, except for the cross bolts.

Mirror 105: Just like it says: a backward mirror image number "105" casting mark commonly found on the driver-side front face of blocks cast at Ford's MCC foundry starting somewhere in the early to mid 1970s. It's usually, but not always, a later-model 390 block with the extra main webbing.

352: The 352 designation is found on the driver-side front face of many of the FE blocks cast at Ford's Dearborn Iron Foundry (DIF) throughout the 1960s. This does not mean you have a 352 engine, or anything else for that matter. Most 390 and 428 engines, as well as many 427s, have this marking.

66-427: This one is often found on the inner valley above the lifters or on the bellhousing face. It tends to get folks really excited for a few minutes, but it means pretty much nothing. It's often found on otherwise normal 390 engines.

C Scratch: This is a good one to find. Found as a freehand letter "C" scratched in the bellhousing area of the block, this is considered a good indicator of the double-webbed 428 CJ block.

A "352" casting in the bellhousing area does not mean that it's a 352 engine. Instead, this is a common casting designator in the FE engines of the era. The 352 casting designation was often found on most 390 and 428 engines as well as many 427s. Also visible are staked oil galley plugs, which are a simple low-cost alternative to screw-in conversion.

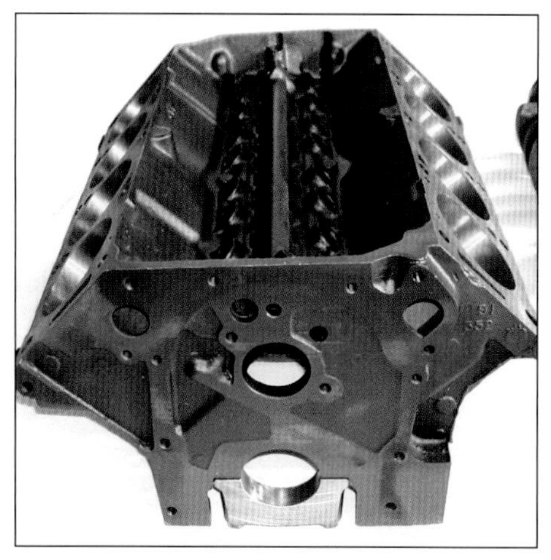

Front of a common FE block showing the "352" casting mark. All FE blocks have bore spacing of 4.63 inches, a deck height of 10.17 inches, and a main journal diameter of 2.749 inches.

The "428" designation is cast in the base of the water jacket as seen through the cylinder head deck-cooling opening. Similar casting numbers are also often found below the center freeze plug. This is a true indicator of the block's 428 casting core.

You're looking at the 428 rear surface. Most 428 blocks have a hand-scratched "A" or "C" in the bellhousing part of the casting. The "A" is generally considered to be a passenger-car 428, while a "C" is a Cobra Jet. There are no scratch marks on this one, but it is a 428!

A "66-427" cast in the bellhousing can be exciting to see, but is often found on otherwise-common 390 engines. Ford used the casting molds on everything with regard to the actual engine block being cast.

A Scratch: Another one that's nice to find. This is the letter "A" free-hand scratched into the bellhousing area casting and normally associated with non-CJ 428 engines.

Inside the Water Jackets: Proof positive of a 428. If you remove the center freeze plug, you can often see the number "428" cast right into the base of the water-jacket core. You can also find similar casting identification by looking straight down through the water opening on the decks where the head gaskets go—you'll need a flashlight.

Casting Numbers, such as C6MA-xx: These numbers are normally found cast upside down below the oil-filter-mounting pad. Unfortunately, they don't really mean all that much. While important for a restoration project, the fact is that Ford used the same casting number across a wide variety of engine sizes and levels. That means that these numbers do not help for identification other than for exclusion. You know that a D4TE (the "D4" indicates 1974 in Ford code) is not going to be a 352, which was discontinued in 1966.

Date Codes: Often, but not always, cast in place alongside the casting number, the date codes tell you when the block was made. Like the casting number, these do not tell you anything about the engine itself other than by exclusion; i.e., a block cast in 1964 is not a Cobra Jet since those started in 1968. Date codes are the "holy grail" for restoration work, but they have limited value for performance efforts.

Cross Bolts: It's most likely a 427, unless a racer or hot rodder has added them sometime in the block's history.

Some scratches were done by hand as an identification method in the foundry, and often look "backward" or crudely formed. They sometimes can provide a means of identification although many marks were randomly used and don't provide any actual ID information. A "C" scratch on back of a 428 is considered a good indication of a Cobra Jet block, when combined with date codes in the 1969–1970 timeframe.

Windows in the block! This is an unfortunate but common finding. The rod broke and destroyed the block. A stock rotating assembling can usually handle plenty of horsepower, but the rods are prone to failure after years of hard service. They break in the thin area about 2 inches below the pin. Often, windowed 427 blocks can be repaired and raced again, but this one is broken through the cross-bolt boss, so it is best relegated to street use only now.

The "A" scratch is often found in the bellhousing area on a passenger 428 block. The actual meaning of the letter A is unknown, but it usually appears only on passenger-car 428 blocks.

Screw-in core plugs: Most often indicates a 427, unless a racer or hot rodder has added them sometime in the block's history.

The "Drill Bit Test"

This one test is the single best way to quickly identify an assembled FE block, and credit for it goes to FordFE.com forum member David "Shoe" Schouweiler. You only need the simplest of measuring tools–some drill bits. The following is paraphrased from several of Dave's responses to block identification questions posed on the forum.

Remove the center freeze plug from the side of the engine block. Using common drill bits, then slip the shank portion of the largest possible bit in between the center cylinder cores through the freeze plug opening. The size of this largest drill bit indicates which water-jacket core was used to cast the block.

If you can only fit a 1/8- or 9/64-inch drill bit shank between the cylinders at the largest gap position on the block, and a 10/64-inch bit doesn't fit anywhere, then they are 427 water jackets.

406/428/DIF361/DIF391 blocks allow a 13/64-inch drill bit shank to fit into the gap at the largest position.

MCC361FT/MCC391FT blocks (MCC = "mirror 105" marking) allow a 14/64-inch bit to fit between the cores.

Regular 360/390/410 blocks hang around the 17/64- to 19/64-inch water-jacket space at the largest position on the block.

These are only approximations, but tend to be close.

Even if you do have the good jackets, be sure to sonic map the cylinders before boring. If the core has shifted, it could cause problems. It is not at all unusual for FE engines to have considerable core shift. And the oft-raced and abused 427 engines seem to have some of the thinnest cylinders. A block with core shift has cylinders that are thicker on one side and thinner on the other. This can leave the cylinder wall too thin after machining, compromising strength and piston-ring seal.

The New Blocks: Aftermarket Offerings

The FE engine's recent renaissance has fostered considerable interest from the automotive racing aftermarket. Among the new FE parts now available are several brand-new engine blocks from, in no particular order, Dove, Genesis, Pond, and Shelby.

Most of my personal experience has been with the Genesis and Pond offerings. Thus, they are covered here in detail. Any of these blocks are considerably stronger than any factory parts.

The factory 427 engines can be identified by their unique cross-bolted main caps, which have small spacers. These spacers are unique to their position and reside between the oil pan rail and the cap. The aftermarket block suppliers have taken this a step further by integrating the spacers into the main caps themselves. On the Genesis blocks, the spacers have a protruding flat section that looks like a "T." Pond blocks extend the entire cap to the side of the pan rail and have a larger flat register alongside the main bearing bore. Pond also adds a dowel pin for positive cap location.

The Genesis, Pond, and OE blocks are rarely found in one location. Having them all in one place makes for some interesting comparisons. Both Genesis and Pond are continuously improving their products, so those shown here will not likely be identical to the one you purchase.

Several FE blocks were used for the comparisons in this chapter. Included in this picture are Genesis blocks in both iron (black; upper left) and aluminum (back row), Pond aluminum blocks (front row), and a factory 427 in blue (right center).

The Genesis iron block has siamesed cylinders; they join together rather than having water flow between them. This allows a bore diameter far larger than any factory FE. At Survival Motorsports, we've overbored 4.375 inches frequently, compared to the original-equipment (OE) parts, which become marginal at anything beyond 4.270 to 4.280 inches. The Genesis and Pond aluminum blocks use pressed-in-place cylinder liners and are good to roughly 4.310-inch bore diameters. Any build targeting serious high horsepower would be well served by starting with an aftermarket block.

The Genesis block looks virtually identical to a factory 427 on the outside, with the same screw-in core plugs and casting contours. Even a sharp eye cannot tell this from an original. The Pond block looks very similar with minor changes—larger cooling passages make for a smoother exterior, and the use of CNC O-ring sealed plugs tell the knowledgeable observer that this is not a stock piece.

We Have a Block–Now What?

If you are starting with a used block, the very first step is to qualify your candidate and make sure it's sound for the build-up process. Each step in this process is intended to eliminate any problems before they are found—before wasting money on an unusable block.

Do a rough bore measurement. With very rare exception the FE family is noted for fairly thin cylinder-wall thicknesses, so it is best to keep the overboring at a minimum. On any 390 or 428 block, I prefer to stay at .030- or .040-inch over when feasible, and to avoid going much larger unless it's required.

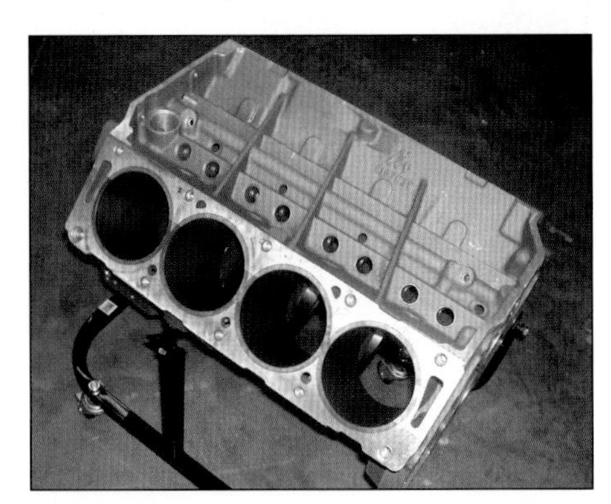

A Genesis aluminum block with lifter bore bushings, which are used to optimize bore geometry in a solid-roller cam race engine.

Since 390 blocks are relatively inexpensive, it's usually easy to find a standard bore of .030-inch over (4.080) one to start from. Unless you are in a "numbers matching" situation, a rough or larger overbored 390 block should be simply replaced. Some online marketers make a business of boring 390 blocks .080-inch over and calling them standard-bore 428s. This should be avoided in the vast majority of cases (more on this later).

The side of this Genesis aluminum block shows the physical similarity to a factory casting. Other than the alloy, it is externally identical to an original part.

There have been a few versions of Genesis block main caps over the past several years, but they have always used a one-piece, integral cross-bolt design, as opposed to the factory-style spacers. Billet-steel caps are now an available option for very-high-powered race applications.

On 428 blocks, the game is a bit different because they are much harder to come by. I have been able to use a program from Diamond pistons to go larger in very small increments, .025-, .035-, or .045-inch over. Couple this with the readily available .030- and .040-inch-over pistons and you have lot of options. Again, the best strategy by far is to keep bore diameter to a minimum. You lose a lot more power due to poor cylinder sealing than you can possibly gain in cubic inches.

This is the Genesis iron-block bottom end. Compare it to the factory block photos on page 16, and you can see the thicker bulkheads and reinforced casting on the Genesis block. This reinforcement does add some weight and therefore the Genesis weighs at least 30 pounds more than the stock block.

On the most valuable 427s, it's best to stay as close to the original

With the Genesis iron block, the lifter-valley cross ribs are included, but not needed. They are required for an aluminum block, so Genesis made them part of the casting mold. In addition, the block is machined to accept hydraulic lifters, so it can run all Ford OEM and aftermarket valvetrain parts.

The Genesis 427 side-oiler iron block looks much like the original that inspired it, but it includes numerous features that make it much stronger. For example, it has thicker, siamese-cast cylinder walls permitting larger bores and thicker decks. In addition, the reinforced cam tunnel permits modification for use of 2.500 roller-style bearings if desired.

bore as possible. These are the thinnest-wall FE blocks, and yet they see the hardest use. It is very common for the first oversize to be 4.250—.017-inch over. After that, builders often use a custom piston to keep material removal to the bare minimum. In the context of a 427 build, the added expense of the custom piston is modest.

Now do a quick physical inspection. Look for the obvious. FE engines are noted for cracks in the deck running into the head bolt holes, and they sometimes crack in the main oiling galleys running up to the cam. I've seen some that had cracks at the root of the main bolts, and plenty with freeze damage in the lifter valley area. Stripped-out head bolt holes are not unusual. Welded-shut "windows" from thrown rods in the pan rail area are common in old race 427s. Any of these probably relegates a 390 to the scrap pile, but even a badly broken 427 gets fixed almost without exception.

Cleanup Time

If you have found a block with a decently modest bore size and no visible major flaws, it's time to get started. My shop preference is a bake/media blast/wash cleaning process that leaves an old block looking like a new casting. Other shops use other methods. No matter how, you must get the block as clean as possible. Run a cleaning tap into all of the threaded holes (this is not the same as a threading tap, which removes more material). Once 40 years of grime and scale are removed you might see flaws that were previously invisible. You need the water jackets really clean in order to perform the next step—a sonic test.

Sonic Testing

A sonic test is not mandatory, but is a very good investment (around $100) that can help define your build strategy for a marginal block. An inexpensive 390 block that is at standard bore or .030-inch over won't likely get a sonic test, but it should be budgeted as part of every expensive build or for an older 427.

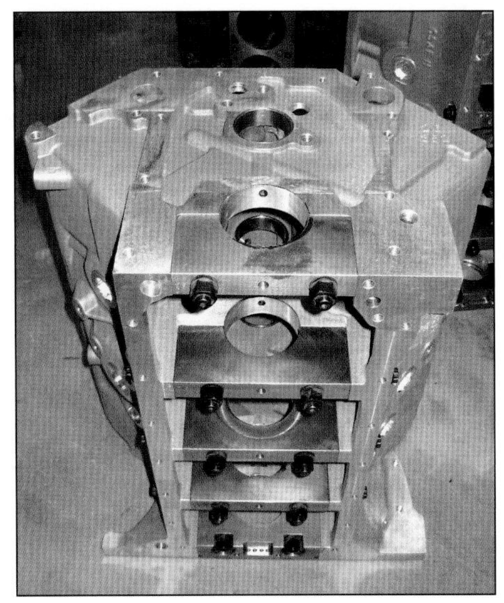

In this photo, you see the bottom end of a Pond block in detail. Note the different main-cap design, which goes all the way to the pan rail. Pond FE 427 aluminum blocks can handle more than 1,300 hp and are substantially stronger than the original cast-iron Ford FE block. Pond aluminum blocks include billet caps and dowel pins, which are very desirable features.

Pond block side view shows the altered side casting and billet core plugs. These blocks still look reasonably similar to a factory block, but they are visibly identifiable as aftermarket parts.

Done properly, a sonic test delivers a map of each cylinder at several spots around the bore and in multiple positions up and down the cylinder. A dozen measurements per cylinder are common. The sonic tester uses sound waves to get a thickness reading and requires the user to properly calibrate the device and to use the right contact probe. It reads rust and scale as well as iron; that's why things have to be cleaned first.

If you have done a drill-bit test earlier in your qualification process, you have the data to cross check the sonic measurements for accuracy. The FE engine has 4.630-inch bore centers. If the cylinder bores are at 4.250 inches, and there is .140 inch (9/64 inch) measured between the cylinder cores, it would be physically impossible for the added wall thicknesses of the two adjoining cylinders to exceed .240 inch. If the readings show otherwise—like .135 inch and .130 inch—you know they are wrong and need to be redone.

A sonic reading that shows a wall thickness below .100 inch is cause for some concern. Any reading below .090 inch is getting to the low safe limits. The location of the thin spots should determine whether or not to use the block. A thin wall low on the cylinder is not going to see as much cylinder pressure or heat. However, a thin section in the middle of the cylinder is an opportunity for failure because the heat levels are the highest, flex is the greatest, and support the lowest. And likewise, those online auction blocks with "standard-bore 428s" created from .080-inch over 390 blocks are going to fail in this same area. On the high-dollar 427 blocks that come in thin at the bottom of the bore, I often use a short pour of block filler in the water jackets (an inch or so) before machining commences, to provide some extra support to the bores.

Valve-to-Cylinder-Bore Clearance

This is something that needs to be checked before engine machining, but almost never is. You won't have a problem with most FE combinations, but a couple can "catch you." The exhaust valve on an FE is closer to the cylinder bore than the intake, so a larger exhaust—like 1.710 inch or bigger—hits the bore on a 390-based engine. Some of the more uncommon 427 heads (tunnel ports and high-risers) had 2.250-inch intakes and 1.750 inches installed; these sometimes hit on a 428 or a standard-bore 427. In any case, the bores can be notched for clearance as long as you keep the notch above the "ring belt" where the piston rings travel.

Machining the Block

After cleaning, checking, and qualifying the chosen block, the machining process can finally begin. Because it is 30 to 40 years old, it's best to assume that your block is going to need the full treatment, which includes boring, honing, decking, and line honing. A super-budget 390 build might get away with just a hone job, but this book is about "Max Performance."

Every shop is going to have its own sequence of doing things, and it can change based on the type of build being performed. One example is the decking operation. While a street engine gets a straightforward parallel deck job based on getting both sides even and square, a race block gets decked after test installing the rotating components and measuring actual deck clearance against the desired value. What matters is being certain that each operation is done properly and is not adversely affected by subsequent work.

Bore and Hone—Cylinder Prep

For this project, cylinder prep is oriented around the use of the exceedingly popular plasma-moly facing on a ductile-iron top-ring combination. These parameters have been repeatedly proven in countless street, track, and production applications. While it's entirely possible for a certain unique finish to have a benefit in a certain unique application, the honing techniques, measurements, and processes described will get you very close every time.

Boring and honing is best done with a torque plate. While plenty of engines are assembled and run fine without one, this is one area where added effort and expense show real benefits. Piston-ring seal is a critical factor for effectively building horsepower, and the FE engine with its thin-wall casting, shows significant bore distortion from the head fasteners. When installing the torque plate, be certain to use the same torque values, type of fastener, and head gasket that will be installed in the engine because each type loads and distorts the cylinders a bit differently. Head bolts tend to pull "up" on the block's upper threads while studs more evenly distribute tightening stress. Wire ring reinforced-style head gaskets try to bend the head and block around the sealing element, while multi-layer-steel (MLS) gaskets do not have such highly concentrated local loads. Therefore, the MLS gaskets are a better option. Therefore, the MLS gaskets are a better option. Since the whole idea of a torque plate is to get a round cylinder under installed conditions, you should use one to eliminate as many variables as possible.

Rough boring should take you within .010 inch of the desired finished cylinder diameter, and not any closer than .003 inch. This leaves enough material for the honing operation to develop a good surface finish. The OEM vehicle and piston-ring manufacturers have literally written books about cylinder-wall-preparation techniques and topography. With piston-ring seal being critical to performance, I cover it in a general "industry standard" fashion here—with a strong recommendation to consult your ring supplier's documentation for added detail.

The desired cylinder finish has thousands of minute cross-hatched scratches surrounding diamond-shaped plateaus of smooth metal. The concept is that the scratches act as reservoirs for oil—needed for lubrication and sealing—while the smooth plateaus act as the load-bearing and sealing surfaces.

I specify a power hone, such as the Sunnen CV-616, and like to use a traditional vitreous stone for honing, rather than diamond stones. The diamonds are faster and cost less in production use. With the correct control and processes, they can work very well. But diamond stones tend to tear and fracture the metal on a microscopic level; they do not provide as controlled a surface finish in a performance-shop environment. The vitreous stones break down during use and constantly present new sharp cutting edges, delivering a clean and repeatable cylinder finish as a result.

The first sizing pass on my cylinders is done with a fairly coarse (around 280) grit. Within .0005 inch of the desired diameter, several strokes with a 400-grit hone follow so that the highest spots from the first stones are leveled out and plateaus are developed. I finish with a few strokes of a hone-mounted brush, which serves as a cleaner and removes any torn or folded metal. On occasion I may tweak this process by changing grit, speed, or load, or by polishing the finished bore with some 600 wet/dry paper wrapped around the hone, but this basic procedure gets 90+ percent of the engines to where they need to be.

I don't advocate a completely polished or mirror-like cylinder-wall finish. Inadequate scratch depth or area does not give the oil scraped off by the rings any place to go, and the rings can then hydroplane on the resultant thicker oil film. Oil control issues can result, that in turn permit contamination of the combustion area.

Profilometers and fax film are a couple more sophisticated measurement tools used in the high-end racing and developmental environments, and are rarely seen in the average or even above-average shop. However, knowing that they are available can be helpful when confronted with a ring sealing problem or the need to try something new.

A profilometer can be used to generate a series of measurements that quantify the cylinder's surface by defining the average roughness of the surface (rk), the depth of the scratches (rpk), and the volume of the scratches as a percentage of the overall surface area (rvk). The basic roughness number is referenced most often yet is the least informative because it is a simple average—you can get the same number from a very deeply scratched surface as from a shallow scratched surface if the high and low portions are of the same percentage. The volume number is most useful because it quantifies the plateau area against the scratches/voids, which are necessary parts of the cylinder-finish equation.

Fax film is dissolved acetate, applied to the finished cylinder bore. Once dried, the acetate can be carefully peeled off to provide a direct imprint of the finish, which is then examined under a microscope. This non-destructive test can quickly pinpoint torn metal, one-way hone scratches, and other issues that may be otherwise invisible. The fax film shows the difference between a diamond hone and a vitreous hone finish in an instant.

At the proverbial "end of the day," a lot of research and experimentation has proven that the long-accepted honing techniques are actually pretty darn good. Sometimes old ideas and methods were the right ones.

Line Honing or Boring

With a 30- to 40-year-old block, it's best to assume that everything

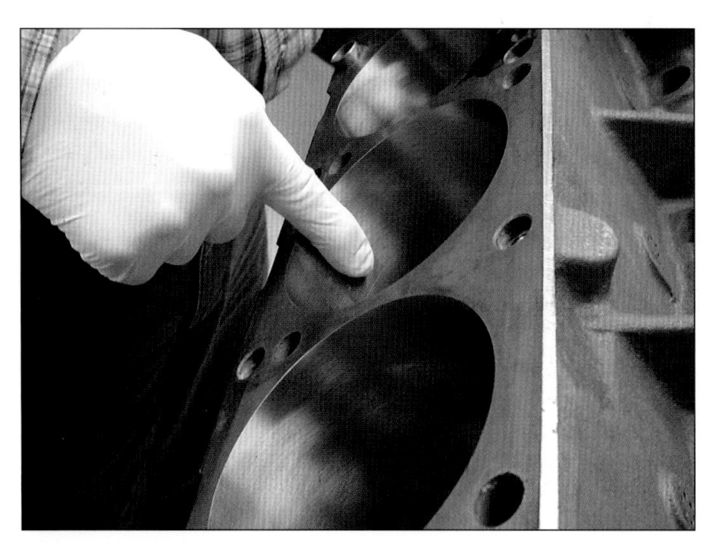

With the fax film process, hold a piece of the supplied plastic strip against cylinder for several minutes, allowing the gel to partially dissolve the plastic, and then to harden.

Applying fax film is a technique used by ring manufacturers and high-end race shops, as a means of checking cylinder wall finish. The process allows you to make a plastic impression of the cylinder surface to verify its condition. The first step is to apply acetone gel to a cleaned cylinder wall.

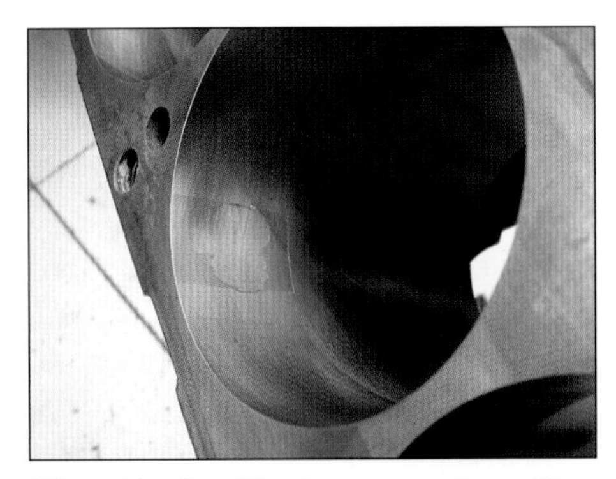

When the fax film is removed, you'll have a perfect plastic impression of the wall finish. Ring manufacturers then examine the plastic strip under a microscope to help solve problems or verify a honing process. They can see characteristics such as the depth of the cross-hatch grooves and the percentage of flat plateau surface versus groove volume. They can also verify torn rather than properly machined metal.

needs service. Line honing—the establishment of straight, round main-bearing bores that have the correct diametrical measurements—is one of those service items. The main bores have a tendency to distort after use, and the proper alignment and diameter are critical factors in bearing durability.

In a similar manner, most new blocks are machined in fixtures that do not permit sizing the main tunnel in a single pass. This means that the machine has to come first from one end of the block for three of the main saddles, and then from the other end for the final pair. Some degree of misalignment is an inherent risk.

The method for checking main bearing tunnel geometry is twofold. First is a check for alignment using a precise dowel rod and feeler gauges to check for variations in the main saddle's position relative to each other. Once this has been determined, I then use a dial bore gauge to measure for the roundness and sizing of each saddle. These dimensions are critical in that they provide for adequate crush and the underlying geometry for the bearings.

"Crush" is a bearing engineering term that best translates to: press fit in a split-bearing installation. This press fit is what actually keeps the bearings in position and prevents "spinning." Those small tangs are only locating devices and are often not present on newer engines. It has become common practice to target the tighter end of the specified range for main saddle diameter with the intention of obtaining the most crush without deforming the bearing itself. Race bearings have stronger steel-backing alloys, which permit installation with greater amounts of crush, and are better suited to this strategy.

The normal method of correcting for improper main saddle dimensions is to mill or grind a small amount of material from the main caps and then use a long mandrel hone to resize the tunnel in a single operation. As with cylinder honing, you should use the same fasteners that will be installed during final assembly.

There are definite limits to the amount of material that can be or should be removed in a honing operation. Large misalignments or working with dissimilar material, such as steel caps on an aluminum block, require line boring. Boring is a more complex operation in terms of setup, and requires less-common equipment, but it is sometimes the only appropriate option.

Whether from repeated line honing or boring, it is possible to remove enough material to impact the center-to-center distances between the cam and crankshaft tunnels. Cloyes offers special "-.005" and "-.010" timing sets for FE engines to address this issue should it arise. Making significant changes in main cap vertical orientation can also become problematic in 427 FE engines due to the main cap cross bolts. These fasteners need to be test fit and the block's cross-bolt holes and spot faces may need added clearance if the caps have been cut by a large amount. Check for adequate clearance in the rear cap for the oil slinger machined into the crankshaft. If you've cut a lot off the cap it can get "tight."

One other thing to check for is a large enough chamfer on the thrust bearing saddle. OEM blocks have a very large 45-degree chamfer where the thrust bends going from the straight shell to the thrust face. Some Genesis blocks have a smaller chamfer, which interferes with some brands of thrust bearing. It's best to check and enlarge that chamfer now.

Machining the Decks

Similar to the other operations, deck machining should be considered a "given" during budgeting. The nominal deck height for all FE Ford blocks is 10.170 inches. Given the age and unknown history of older blocks, it is safe to assume that they have seen service in the past. At the very least, a clean, flat deck surface goes a long way toward ensuring a good head-gasket seal.

With the main saddle line hone finished, it is pretty easy to get an initial reading on deck height using nothing more than a 12-inch dial caliper. Measure from the main bore to the center of the deck at all four corners—front left and right, and rear left and right. Choosing the lowest corner and using the appropriate fixture to mill both decks to that level gets you close enough for the majority of street and bracket racing builds. For these types of engine builds, cylinder-to-cylinder equality is more important than the actual value you arrive at. And it's a good spot to start on a race engine as well, since you are targeting a specific deck clearance, and that lowest point is certainly not going to get any higher. On a serious race engine, you will likely revisit this operation after trial assembly if you need to achieve a particular dimension because there is enough potential for dimensional variation in crankshaft stroke, rod length, piston pin position, or bearing variation to warrant the effort.

When milling the deck, it is important to determine the head gasket you will be using. The popular wire ring gaskets, such as the Fel-Pro 1020, are quite forgiving on surface finish. But the multi-layer-steel (MLS) type of gasket requires a much smoother deck surface to perform properly.

After the milling, it is important to chamfer the head bolt holes, especially if you are using bolts. You should also chamfer the dowel pin positions, as they will be much easier to tap into place without distorting them. I prefer to very lightly break the upper edge on cylinder bores as well. You only need a smooth edge break on these and not a big chamfer—just enough so that rings and fingers don't get caught on an overhanging or sharp edge.

Oiling Modifications

Here is where you get into some controversial territory. Some folks feel that the FE engines have a marginal oiling system that requires a great deal of modification. Others believe that the system is okay with only modest attention to detail required. I tend to fall in with the latter group, making few significant changes on my own engines, but I try to cover some of the alternate approaches because the arguments in favor of the former do have some merit.

Galley Plugs

There seems to be no rhyme or reason for factory use of threaded versus non-threaded plugs. In fact, I've seen both used on the same block. I actually attempt to replace all of the oil galley pressed-in plugs with 1/4-inch NPT plugs. Sometimes you cannot get them all due to core shift and drilling variances. Thin cast iron can be very fragile and some risks are not worth taking, but try to get as many as possible. The pipe tap tells you to use a 7/16-inch drill. I've found that an additional short-entry drilling with a 1/2-inch bit helps get enough threads started for a clean tapping with less side loading and stress to the casting. You need the plugs to go in deep enough, especially the ones in the rear, which can interfere with the flywheel cover plate or in front by the distributor hole.

Be sure to install the hidden plug behind the distributor hole. That one trips up novice and professional alike. It's a safe one to leave as a press-in plug because it cannot go anywhere once the distributor is installed. If you choose to thread it, be certain to check for adequate distributor clearance after the threaded plug is installed. I've seen many engines where this plug interfered with distributor installation or locked up the distributor once installed.

Oil Pump Mounting Area—Pump Outlet

Many of the factory FE blocks have a very small opening passage into the block from the pump. This passage is far smaller than the outlet on the commonly used Melling oil pumps, and has a nearly immediate 90-degree turn leading into the oil filter mounting pad. Mark the opening to match the pump outlet and then use a combination of carbide burrs and cartridge rolls on a die grinder to open up, smooth, and blend this passage. Some builders use a three-fluted drill and open up the galley to 7/16 inch, but I remain more concerned with the shape than the diameter, feeling that a 3/8-inch hole can flow plenty of oil.

Oil Filter Pad—Leading Back into the Block

This is an area where I do not go as far as some other builders. Most blocks have something of a "step" in this passage where the drill size changes. I remove and blend that step, but others drill the entire passage out to a larger size. Similar to the pump outlet, I feel that the galley is adequate once the bumps and burrs are removed.

Oil Passage Alignment

The main bearing oil-feed passages on many FE engines are off-center and significantly out of line with the holes in the bearings. While the reasons may be "lost to the sands of time," it appears to be more a result of manufacturing convenience than any particular design element. The feed holes are drilled straight through the center of the cam bearing journal, and since the cam journals are not in line with the mains, you end up with a variance. The degree of feed-hole impedance varies with brand and type of main bearing. Some have a large feed hole and others a narrow feed slot. The factory 427 engines had the holes chamfered over to line up with the bearing, and this has become common practice with most performance-oriented FE builds. Under extreme race duress, the 390 and 428 engines split through this hole, so it is a good idea to remove as little material as possible in this area. I use the desired bearing as a reference and just blend the edge with a die grinder and a carbide bit.

Restrictions—Lifters and Valvetrain

There was no oiling at all to the lifters in the original 427 Ford engines. This fact led builders converting other FE engines to solid lifters into completely blocking off oil on those too. You never see that done on any other engine, and there is no reason to do so on an FE. You can restrict the lifter oiling on solid or solid-roller engines by using .060-inch drilled set screws, installed into the lifter galley feeds that "V" off from the center of the intake valley. Most blocks can be simply tapped to 3/8 inch in those locations. On the occasional block where this cannot be done, I go without restricting and do so with no regrets whatsoever.

Restricting the oil feeds to the rocker assembly is a good idea on most FE engine builds. These engines seem to put a great deal of oil up to the top end and have lesser drainback provisions. The restrictions are most often installed in the cylinder heads, but the feed holes can also be drilled and tapped for 5/16-inch setscrew restrictions in the cylinder decks if desired. Restriction to .060 inch is common; the Fel-Pro 1020 already has a feed hole with a reduced diameter of .093 inch. On certain Genesis aftermarket blocks, it's important to check oil-feed hole alignment with the Fel-Pro head gasket. They have been known to be off location, and this is by far the best time to identify and correct any such issue. Either drill an extra feed hole in the gasket or use a die grinder to put a shallow transfer groove into the block's deck surface.

Other Passages

Use a small flashlight and gun-cleaning brushes to ensure that all the galleys are clear, clean, and free of burrs and misalignment. I do not advocate drilling out the entire center feed passage to a larger size. It's a big risk with no real benefit, but others disagree.

The front and rear cam bearings have oil holes that need to be in the proper orientation. On side-oiler engines, it's important that the cam bearing side holes line up with the valvetrain feeds to the decks. Using an LED flashlight and looking through the deck feeds, you can see the center of the cam bearing. The hole's position is not as critical in center-oiler engines because they have an annular groove surrounding the three center cam bearings. It's "nice" to have the cam feed hole at around 4 o'clock as viewed from the front, but I've seen them installed every which way without apparent ill effect. You can add grooves connecting the oil passages in the cam bearing tunnels of a side-oiler to permit the use of common FE cam bearings.

Other Details

I do not like to paint the insides of my blocks. Some folks use Glyptal or similar products, but in my opinion, paint inside an engine is a risk with no real reward. Paint is not going to significantly affect oil drainback. For proof, pour some oil on a rough casting and see how fast it runs off. It'll hit the floor before you can catch it. Paint inside the engine isn't effective for capturing dirt, either, because significant dirt or debris under that coating causes the paint to come off.

I spend time smoothing rough edges and removing casting flash. But I do not polish the inside of a block. I have never seen a block crack initiated from a rough spot in a casting, but have seen plenty of damage from grinding dust and dirt. Why add more of that?

Be sure to chase all the threads, both internal and external, before putting the block into the assembly room. Few things are more frustrating than finding a broken bolt or a stripped hole in a minor accessory position after much of the assembly work is done.

Many early blocks have an intake alignment dowel pin installed on the front intake rail. Most aftermarket intakes do not accommodate the pin, and I remove them as a part of normal prep.

I do paint the outside of blocks before assembly. Right after washing them for the last time, I mask them off and spray the color. I seem to get a nicer finish and better adhesion before putting assembly lube and such on the casting.

And then it's time for the other parts.

CHAPTER 3

CRANKSHAFTS

The Ford FE engine family shares common key crankshaft dimensions, making it fairly easy to interchange from one displacement to another and allowing a broad selection of strokes and materials (OEM and aftermarket).

OEM Choices

In the original-equipment world, a wide range of crankshaft strokes was available over the 20-year span of FE production (1958–1978), from the short-lived 3.300-inch-stroke 332-ci engine through the 3.980-inch-stroke 428. Any of these can be physically installed into any FE block providing that coordinating rods and pistons are used.

Original-equipment crankshafts were made from either cast-nodular iron or forged steel. The nodular-iron cranks have proven to be extremely durable and are found in the vast majority of production engines. A steel crankshaft is inherently superior—especially in severe service, in which it is exposed to extremes in terms of power or cycle fatigue. A road-race engine is exposed to extreme cycle fatigue because it is subjected to long periods of heavy throttle and transitional loading. A steel crank is ideally suited for this application. On the other hand, a cast-nodular-iron crank is better for the budget-constrained street-oriented builds and many drag-race applications because these cranks are typically subjected to full-throttle operation for shorter periods of time.

An imported aftermarket steel 4.250-inch FE stroker crank gives you a lot of cubic inches for a modest cost, and it's able to support at least 750 hp or 7,000 rpm. While certainly good enough for most street/strip applications, this is not comparable to a genuine race billet crank in terms of strength or accuracy.

Steel crankshafts were installed in the always-rare 427 performance engines with their 3.780-inch stroke and in medium-duty truck 361 and 391 engines. The limited supply of factory 427 steel cranks in repairable condition drove their cost beyond the reach of many engine builders several years ago. There are a scant few NASCAR 427 steel cranks that use a wider connecting rod and bearing than common OEM equipment. With these supporting parts being nearly unobtainable, the NASCAR cranks are best left in the hands of collectors who have a demonstrated need for such parts.

It has become quite common to modify the 3.780-inch-stroke steel 391 truck crankshaft, converting it into a performance piece. The truck crankshaft has a larger-diameter snout where the damper mounts, as well as a different flywheel mounting

Factory FE steel cranks contain either press-in rod core plugs (left) or screw-in plugs (right). The running change from press to screw design was made with the same part number. There is no real advantage to one design over the other.

The press-in steel crankshaft plugs have a Tru-arc-style snap ring as a retainer, and it's wise to use new snap rings after service. If a plug doesn't have a retainer, it can come apart at high RPM and cause catastrophic damage.

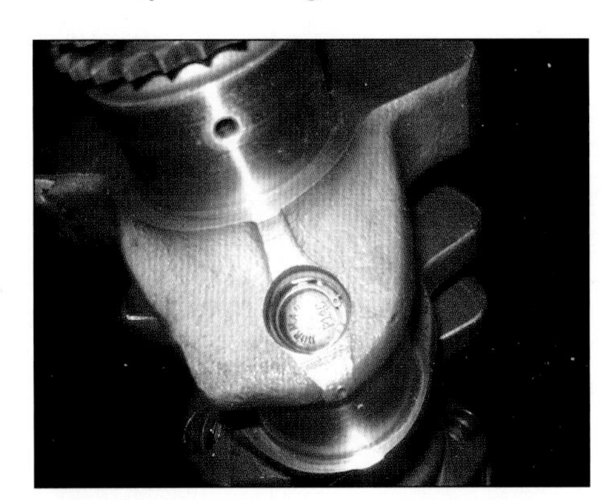

Here is a counterweight plug in a factory steel crank. These plugs should be removed during cleaning because debris often finds its way into the passage behind them.

flange and a counterweight combination designed for external balance. The pilot-bushing (or converter snout) hole in a steel truck crankshaft is larger in diameter as well, mandating a custom pilot or converter bushing for use in passenger-car applications. The bushing must have the same ID, but the OD is larger. The required modifications are time consuming, but a skilled crankshaft shop can handle them. The crankshaft nose needs to be cut down to the standard 1.375-inch FE diameter, the flywheel mounting surface is shortened by about .060 inch, and the counterweights need to be cut down before balancing is possible. The supply of usable 391 cores has also dwindled in recent years, increasing the cost of the finished product, but this remains a viable option for those needing a steel crank within the stock stroke ranges at half the cost of a custom billet.

By far the most common factory offerings were the 3.500-inch-stroke cast crankshafts found in 352 and

The 427 was the only FE passenger-car engine equipped with a steel crank. The 332, 352, 361, and 390 all came equipped with cast-iron cranks. The factory steel cranks have a dollar-sign forging mark, which has become something of a legend considering the market price of these rare parts.

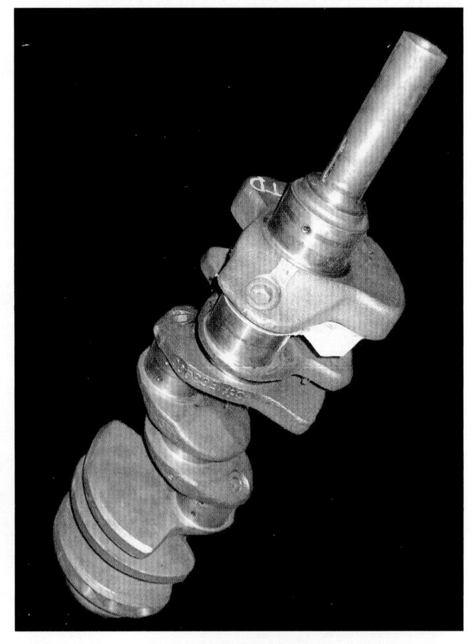

Screw-in-plug steel crankshafts are otherwise identical to their press-in counterparts, and there is no real advantage to one over the other. An unsubstantiated rumor is that the change was made after some press-in cranks had the holes drilled too large for the plugs to fit.

360 engines and the 3.780-inch-stroke cast units common to 390s. The highly sought-after 3.980-inch-stroke cranks were only used in 428 applications and for two years (1966–1967) in Mercury 410 engines. A few of the 428 crankshafts have additional center counterweights, and were specified for vehicles equipped with the Super Cobra Jet (the "Drag Pack" option). Your odds of finding one of them in a normal swap meet or junkyard are slim.

Identification

Once you've seen an FE crankshaft, it's pretty easy to pick out another one among any collection of parts. They have a unique long front snout similar to that of the 429/460 parts, but are much smaller in main journal and counterweight size.

It's a fair bet that the average FE crankshaft you find in a parts pile or swap booth will be a cast 390 or 360 piece. If you are really lucky, you may find a 428 crank. A steel truck crankshaft occasionally shows up, but most often they have been turned .030 or .040 under. A serviceable steel 427 crank is extremely difficult to find outside of Ford specialty retailers or among dedicated racers.

It's fairly easy to tell a cast crankshaft from a steel one. The cast crank has a very thin parting line where the two halves of the mold meet one another. In comparison, the steel crankshaft has a thicker parting line—usually between 1/4 and 1/2 inch wide.

Ford forges or casts numbers on the counterweights of the FE cranks which may simplify identification. As is the case in most FE parts, the numbers are helpful but not definitive. Often they are useful for exclusion; if a crank is stamped "2U" it's obviously not a 428 part, but the absence of any stamping number does not preclude it from being one.

Here are some common casting marks; it's by no means a comprehensive list:

- The cast 360 often has a 2T or 2TA
- The cast 390 often has a 2U or 2UA
- The cast 428 has a 1U, 1UA, or 1UB
- The cast 428 SCJ has a 1UA or 1UB
- The steel 427 may have a somewhat appropriate "$" sign

These parts are now 30 to 40 years old and have often seen a vast amount of modification, alteration, repair, and service. By far the best way to identify the crank is to measure the actual stroke using V-blocks or even an engine block with a couple main bearings set in it. The stroke variances are large enough so that you don't have to be perfectly accurate for basic ID; even a ruler does the job. It is around 3.5 inches for the 360, just over 3.75 inches on the 390 cranks, and nearly 4 inches for the 428 version. The extra center counterweights on a 428 SCJ crank are obvious, as is the large 1.750-inch-diameter snout of an unaltered steel truck crank.

Typical FE steel crank renumbering is located on this area of the crankshaft counterweight. The factory ground down the casting number and stamped these numbers into the crank. The simplistic appearance is often mistaken for a machine-shop operation done during prior service work.

Aftermarket Offerings

The past few years have seen an explosion of aftermarket products for the previously neglected FE engine family. Currently, many crankshaft options are readily available. In most cases, these cranks are imported products with a broad range of quality. However, they are also the foundation for economical assembly of some powerful but previously uncommon combinations.

Scat was first to the aftermarket with low-cost stroker crankshafts for the FE. And among the imported products, it seems to stand out in terms of having very consistent dimensional quality. At this time, it offers nodular iron cast cranks in strokes of 3.980 inches (essentially a 428 replacement item), 4.125 inches, and 4.250 inches.

The Scat 4.250-stroke FE crank is a durable, ductile iron piece, which is perfectly suitable for any street use, as well as moderate race applications. These are easily dropped into a stock FE block and no clearance grinding is required. I've run them at more than 700 hp with no issues.

Constructed of E4340 aircraft alloy, the Scat FE 4.375 billet crankshaft is intended for high-powered race use. These cost more than $2,500, but are the right item for an 800-hp, 8,000-rpm race engine or a road-race application. These cranks are precision ground, heat treated and nitrided to cope with all-out racing conditions.

The Scat 3.980-stroke crank is designed for internal balancing, which differs from the factory piece. However, other critical dimensions, such as main and rod journal diameters, are the same as the original Ford parts. This crank is a good choice for upgrading or service while using existing OE dimension rods, but the flexplate or flywheel needs to be a 390 part; original 428 parts have balance weights on them and will not work correctly.

The 4.125- and 4.250-inch-stroke crankshafts use 2.200-inch-diameter rod journals. They are designed to be run with common big-block Chevrolet bearings and connecting rods—most often in lengths of 6.700 or 6.800 inches.

In four years of testing and abuse, I have yet to break one of these cast 4.250-inch cranks—at power levels exceeding 750 hp. Intuitively, this has to be approaching the design limits of the material, though, and an alternate should be considered at that power level.

Also available are imported forged crankshafts from other manufacturers—currently in strokes of 4.125, 4.250, and 4.375 inches. These should, by virtue of the claimed 4340 forged-steel alloy, be a physically stronger option for higher-powered or road-race applications. These do not have the same level of machining quality, though, and should be thoroughly inspected and corrected prior to installation.

The top of the FE crankshaft food chain is a billet piece. Available from a number of suppliers including Scat, Crower, and Moldex, a billet crank is cut from a huge piece of bar stock, and can take on almost any stroke and design. These are normally made to extremely high-quality standards, and come with a matching price tag.

Dimensional and Physical Inspection

The most important descriptors for any crankshaft (OEM or aftermarket) are "solid," "straight," and "round." These terms apply to the crank as a whole as well as to each individual journal.

Checking the crank for straightness can be done either on a set of V-blocks or, lacking those, you can use the block itself by cradling the crank with bearing shells in the front and rear journals only. Use a dial micrometer on the center journal to see whether it's straight or not. Try moving a bearing from one end to the center and rechecking at the end journal. While the factory-accepted tolerance per the book may be higher, anything beyond a few "tenths" is a potential problem if clearances under .0030 inch are expected.

The same holds true for main and rod journals. These need to be checked in multiple places around each journal, and on at least a few spots front to rear. It is common to find a rod (or main) journal that has significant diameter differences forming either a tapered or barreled shape. Once again, any variance beyond a few ten thousandths of an inch can lead to early bearing wear or failure. Out-of-tolerance journals are common in used factory parts as well as brand-new items. You need to check. Your crankshaft machinist can grind to the next undersize if needed to straighten out a lot of these variances—a pretty common occurrence on even new imported crankshafts.

There are additional dimensional specifications for rod journal width and main thrust journal width. Rod journal width, along with the dimensions of the chosen connecting rods, determines the rod side clearance. This is a more forgiving clearance with a broad acceptable range (.008 to .020 inch is common), but still something that needs to be validated during assembly. Thrust journal width can be compared to specs while actual thrust clearance, a critical dimension, cannot be determined until the crank is installed in the block with the chosen bearing.

Cranks should be checked for cracks using Magnaflux equipment. While a good idea for any part, this is particularly important when reusing a cast crank because it is very common to find fatigue cracks around the rod journal radii alongside the counterweights. Sometimes these are surface flaws that can be removed by grinding down to the next undersize—sometimes not.

Throw index and stroke variance are less often checked in the typical home engine build, but the latter, in particular, is very important to determine. If you're not paying attention, it can "catch you," and the consequences are catastrophic. A lot of folks target a near-zero deck clearance on their engine project—simply adding the nominal dimension of connecting rods, pistons, and crank stroke—and then machine the block decks to the combined value. Differences in stroke measured from journal to journal of .001 inch or more are not unusual, and I've seen more than .005 inch on some import crankshafts. A bit too much will impact compression ratio, piston-to-valve clearance, and piston-to-deck clearance.

Before you pronounce your crankshaft candidate a "winner," you still need to take a quick look at the snout diameter, the condition of the keyway slot, and the threads for the damper bolt. While often repairable, damage in these areas often renders a crank financially unsound as a prospect.

Connecting Rods

Connecting rods are perhaps the most critical short-block component in terms of load and the impact of a failed part. Rods do not really fail from horsepower. Compressive loads are rarely a problem for a decent modern connecting rod, and all FE connecting rods (both original and aftermarket) are forged steel. Commonly, the rods fail from either cyclic fatigue or overload stress. However, rods must not be the weakest link in the rotating assembly, so you need to select rods that are compatible with the engine combination.

The stress loads come from the weight of the piston and the amount of RPM the rod is subjected to. Higher RPM and/or heavier pistons take their toll on any rod. Cyclic fatigue results from the additive impact of many such loads over time. Every time the rod is put into an elastic state from load it reacts metallurgically, and

A C3AE-C factory connecting rod was abundantly used in FE engines. It is very common to see a variety of rods in an FE pulled from a yard or older vehicle. Large-scale rebuilders often considered most FE rods to be interchangeable and routinely mixed them in service.

The C3AE-B FE connecting rod is visually different from the earlier part. As you check through FE rods, remember that the number on the part is a casting or forging number, not the actual part number found in a Ford catalog. Ford often machined many different parts from the same forging blank.

The C6AE-C, a LeMans-style connecting rod, features a beefy 7/16-inch-diameter cap screw and large-beam cross sections. The LeMans rods were built for high-performance duty in the 427 and 428 SCJ engines. There are minor differences between the various factory iterations, but there are also lots of similarities.

these minute changes in metallurgy accumulate over time and lead to fatigue. This is the reason that a road-race application requires a better-quality rod than a similar horsepower drag car; it's the number of cycles under high RPM.

Factory Rods

With the single exception of the slightly longer and not particularly desirable 352 parts, all FE-family connecting rods share a common, nominal center-to-center length of 6.488 inches—usually just specified as 6.490 inches. They also all have a median big end housing bore of 2.5911 inches and are designed to work with a .975-inch pin diameter. All FE Ford piston pins are of the floating design, and a bronze bushing in the small end of the rod supports the pin.

The standard service and high-performance factory FE connecting rods can be divided into three general groupings. Over the 20-year span of production, there were a number of forging numbers used and some of these were used with no apparent rhyme or reason. So, as with many other numbers on FE parts, the forging number is useful but not truly definitive for identification.

The most common passenger car and truck rods used a 3/8-inch rod bolt and nut for big-end attachment. These are suitable for stock rebuilds and mild-performance applications when upgraded with ARP replacement fasteners and proper reconditioning.

The heavy-duty rods as found in the 428 Cobra Jet engines are less common. These use essentially the same forging blanks, but have a larger-diameter 13/32-inch rod bolt and nut. It is worth noting that the ARP replacement fastener for these rods has the larger shank diameter but still only a 3/8-inch threaded section. These are likely to be stronger in service if (a big "if") you assume that the rod bolt shank diameter is the weak point in the package.

High-performance 427 engines, as well as the 428 Super Cobra Jet, received the "LeMans" rods. There were a few variations on these, but they all shared a much larger and beefier design along with capscrews and locating dowels instead of the

The C7AE-B factory connecting rod is fairly common. It's important to note that the factory numbers on rods and other items are casting or forging numbers, and can be found in numerous machined variations of that casting. A properly conditioned set of these rods can cope with 450 hp, but the same service life and fatigue stress issues apply. By the time you've inspected them, purchased quality fasteners, reconditioned the big ends, and resized the small ends with new bushings, you're nearly at the same cost as a new replacement rod set.

This LeMans capscrew connecting rod has casting number C6AE-E. The fairly rare capscrew rods are instantly identifiable, and these are not going to be found in any common passenger-car or truck FE engine. Because this is a high-performance rod, a set in good shape can support 700 hp or 7,000 rpm. The issue is not the ultimate strength of the rods, but the fatigue life. Most factory FE rods are now 40 years old, and can have internal stress issues that do not show up with a Magnaflux or X-ray.

This LeMans-style capscrew connecting rod has an SK37057 forging number. Generally speaking, an "SK" number indicates a prototype or non-production item. Sometimes the SK parts are noticeably different than their production brethren. Other times they are pre-release parts used in testing and development.

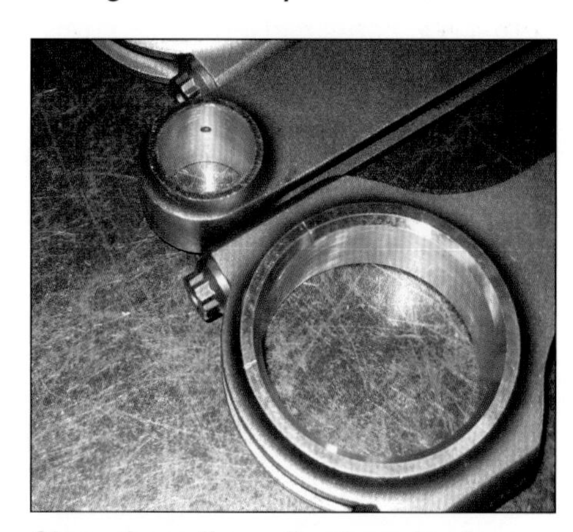

Since they all use floating pins from the factory, bronze pin bushings are used in all FE engines. Some large-scale rebuilders serviced commercial FE engines by removing the bushing and using a larger-diameter press-fit piston pin. Those replacement pins, pistons, and rods have no place in a performance build.

more common nut and bolt fasteners. Despite an undeserved reputation for early bolt failure, these are actually very robust parts and are the obvious design forefathers for many of the current aftermarket rod offerings. Replacement bolts are readily available, but the dowel sleeves are problematic at the time of this writing. The NASCAR version of the capscrew rod is wider and heavier, and requires a matching and equally rare crankshaft.

The biggest problem with any of the factory rods is a simple one of age. With 30 or more years of undocumented service, of unknown severity, many of these are simply getting worn out. Even a thorough physical inspection and good reconditioning techniques do not uncover metal fatigue. Fatigued rods often fail in the same area, breaking about 2 inches below the pin. The cross section of the factory FE rod is surprisingly thin in that area, and it flexes (and breaks) like a repeatedly bent coat hanger after an unknowable number of cycles and loads. It usually happens to the guy who claims, "I've gone xxx number of years with stock rods...you don't need anything else." He is correct in a certain sense. The factory rod could withstand a real beating in 1968 and 1978, and in 1988, when it had better integrity, but it might break instead of bending in 2009. You just don't know when.

Aftermarket Rods

There is a wide variety of aftermarket connecting rods available for the FE engine now. The general breakdown revolves around price point and design. Prices range from less than $300 to more than $1,500 for steel rods that have similar descriptions but very different quality. The designs are either "I" beam or "H" beam with arguments for and against each at every level of the sport.

The replacements for factory-dimension rods are comparatively limited, while the rods used in common stroker kits are based on popular 6.700- or 6.800-, 2.200-inch journal big-block Chevrolet parts and thus have every imaginable option and source.

With notably rare exception, all aftermarket rods come with some sort of ARP fastener. Excellent products combined with excellent marketing have virtually made this a "must-have" feature for rod sales.

The lowest-cost connecting rods are all imported parts sold in unmarked white boxes under a plethora of house brands. The cheapest of these should best be viewed as a replacement for stock rods and little more. With questionable metallurgy and poor dimensional integrity, a really cheap connecting rod is not a good place to save a couple dollars. Reconditioned stock rods are often a safer bet in a stock rebuild because, in many cases, they're dimensionally superior. Scat and Eagle rods are found on the next tier up. At a price point around $500 per set, these are still imported but have considerably higher standards for quality and machining. In this range the H-beam design seems markedly superior, not due to any particular

The Chevy big-block-based I-Beam Scat connecting rod is a good option for building a moderately priced stroker engine. The Scat rods are steel forgings with decent machining and ARP 8740 fasteners.

A Scat 6.700-inch H-beam rod is a solid choice for mid-level stroker engines of up to 750 or 800 hp. It uses a 2.200-inch-diameter Chevy big-block rod bearing and can be upgraded with a stronger ARP 2000 bolt for added clamping force.

engineering advantage, but because of a nicer fit and finish.

With the addition of the optional ARP 2000 bolts, I've run the Scat H-beam parts at power levels beyond 750 hp and 7,000 rpm without incident, and feel confident in recommending them for most high-performance street cars as well as moderate drag-race applications.

At the upper end of the FE rod spectrum are true premium rod sets, which cost in excess of $1,300 and sometimes much more. These are designed for the most serious drag-race engines, as well as road-race engines that produce up to 800 hp. Manufacturers, such as Oliver, Crower, and Carillo, make outstanding quality parts using superior metals and far better dimensional accuracy. While unquestionably superior to the lower-cost parts, the engine builder needs to make the determination of whether the cost penalty outweighs the advantages. In a road-race Cobra or an 800-plus-hp drag car the choice is pretty clear, but less so in a street-oriented car or truck.

Aluminum Rods

Several years ago the aluminum rod was the status quo for a serious drag-race application. The noted advantages of an aluminum rod are reduced weight and a degree of compressive "give" under severe load, which keeps the shock from being transferred to the bearing. The downsides are a greater physical size, dimensional variances under temperature and load conditions, and a shorter cycle life. While still popular in some circles, they are not nearly as dominant as they once were, with the trend for many sportsman drag-race engines going to the more durable steel rod options.

Currently, I am not aware of any FE-specific aluminum rods on the market. There are several billet rod manufacturers who will gladly make some for you if you feel the need. The rods used in most stroker applications are based on the big-block Chevy and are readily available.

With most FE engines destined for sportsman drag racing, Cobra road racing, and street performance, the number of builds that take advantage of an aluminum rod's characteristics are relatively few.

Checking the Rods

Any connecting rod chosen, whether original or aftermarket, should be subject to a thorough inspection before being put into service. Similar to a crankshaft, the rods need to be solid, round, and straight.

Rods should be X-rayed or Magnaflux inspected for cracks or flaws. This is particularly important when reconditioning rods that have already been run for a period of time. Any rod that shows a flaw should be discarded; no common repair is possible.

The rod's big-end bore should be round, straight from one edge to the other, and within diametrical specifications. The diameter is critical because it directly impacts the bearing crush. "Crush," by definition, is a press fit and is the only thing that keeps the bearings from spinning in the rod bore. The little locating tabs do not provide anti-rotation; the crush fit prevents bearing rotation. Newer engines do not even use locating tabs. Equally important is that the bore diameter "repeats" when loosened and re-torqued. Bolt holes that are off from perpendicular, misaligned spot faces, or angled cap surfaces cause the rod cap to shift and change dimensions when assembled. Such issues are common to the lowest-cost rods and expensive to repair, often negating any savings.

The rod's small-end bronze bushing must be straight, round, and sized to the piston pin. With a clearance specification among the tightest in the engine (+/- .001 inch), there is little room for error. The fact that piston manufacturers, including OEMs, do not agree on the exact pin

Scat offers 4340 chromoly H-beam rods as standard-dimension replacements for the FE, as well as the longer Chevy big-block-based stroker parts. These come with ARP 8740 capscrew fasteners. The ARP 2000 bolts are an option for improved clamping force on high-RPM engines.

diameter dimension makes this more critical. When checking pin fit, be sure that any shipping lubricant has been thoroughly removed. Some imported rods have *a lot* of protective anti-corrosive goo on them, and it can really mess up your measurements.

Obviously, a connecting rod needs to be straight in all planes and the pin bore needs to be aligned with the big-end bore. If they're not straight and in line, toss the rod.

That pebbled surface on stock rods is from shot peening, which adds surface hardness and resists stress-riser formation. If you choose to polish the beams on some stock rods, you should have them re-shot peened to restore that surface. Of course, this adds cost, further justifying the purchase of aftermarket rods.

Engine Bearings

Engine bearings are a critical element in any engine, and the FE is no different. The main bearings are fed oil from the galleys in the block. The oil is transferred through grooves in each bearing that in turn directs that oil to the rod bearings through passages in the crankshaft.

These grooves in the bearing can be only in the upper shell, or go all the way around the inside diameter (full groove). Our preference is for the most current design from Federal-Mogul, which uses a groove configuration going 3/4 of the way around the inside diameter. The 3/4 groove bearing allows more surface area to remain on the highly loaded lower portion of the bearing, while still providing more oil to the rod bearings than does a 1/2 groove design.

Some of the inexpensive rebuilder-quality main bearings use the same shell top and bottom—putting a big oil hole right at the point of the highest load. Avoid these.

The thrust bearing in an FE is on the number-3 main. Folding a bearing into the proper shape during the manufacturing process forms the thrust bearings. If the bearing is folded too far, or not far enough, it will not have proper thrust contact. This is the reason for working the crankshaft back and forth a few times during assembly; it helps to form the bearing and correct variances in thrust-face angles.

Rod bearings in performance applications have either a wide chamfer on one edge or a narrowed design. This provides necessary clearance for the larger .125-inch radius found on high-performance crankshaft journals. Without the proper clearance, you may find that the rods bind up when assembled and the bearings wear prematurely from edge loading and contact.

Race bearings are made from higher-quality materials than passenger-car bearings. The difference in cost is modest, and I always specify a race bearing in any build. Contrary to some beliefs, a race bearing is not harder on crankshafts. The only downside to running a race bearing on the street is that it is less tolerant of debris. Frequent oil changes are the norm for a performance build, though, so this is really a non-issue.

A brand-name race bearing has a stronger steel backing, allowing a higher degree of crush to resist spinning and improve heat transfer. Race bearings also have greater eccentricity, which prevents crankshaft contact in cases of rod or main cap distortion.

Do not use abrasives on a bearing if at all avoidable. They get pushed into the lining material and wear on the crankshaft. If you must clean or burnish the bearing, you can try some WD-40 on a piece of brown paper from a shopping bag. It cleans the surface without abrasives.

Assembly Tips

ARP recommends repeated tightening and loosening cycles for rod bolts before putting them into service. This serves to burnish the threads and provides smoother thread engagement and pull up to torque. You will get an adequate number of cycles when assembling and disassembling rods for cleaning and checking clearances. Be careful selecting products used to clean the fastener threads between tightening cycles; chlorinated solvents, such as brake cleaner, are corrosive to certain bolt alloys.

Most rod and fastener manufacturers prefer to use bolt stretch, rather than torque readings to determine adequate fastener tightness. While functionally superior, it can be difficult in some engines to measure stretch with the rod in the block. I often stretch the bolt to spec on the workbench and record the torque value needed to reach that point using that torque spec for final assembly.

When installing rods into the engine, remember that the rod's big-end bore has a large chamfer on one side and a small chamfer on the opposite side. The larger chamfer always faces the crank counterweight and the small chamfers face each other. Mixing this up usually puts the bearing into contact with the crank journal radius causing rapid wear, if you are even able to turn the engine over at all.

PISTONS AND PISTON RINGS

A typical FE flat-top forged piston is shown. Most non-race FE pistons have symmetrical valve pockets, so the same piston design can be installed in all eight cylinders.

In the realm of piston selection, there are a series of basic choices that need to be made first, followed by a lengthy list of detail choices that define the best part for the task. Or you can simply choose a good-quality forged piston and be done with it. But that would be too easy.

Material and Design

The first part of the selection process is material choice, where you have to choose both the manufacturing process and the alloy. Pistons can be either cast or forged. Virtually all pistons today are made from aluminum, but there are several alloys to choose from based on the manufacturing process and the expected usage.

I should quickly define the difference between the piston manufacturing process, the metallurgy, and the design features. Each of these is interdependent with the others and have a significant impact on piston performance. But they are each different subjects, and you cannot make blanket statements about one process or alloy being better than another one without the proper context. Misleading advertising from various manufacturers has further clouded this issue.

Manufacturing Process–Cast or Forged

Cast pistons are made from molten aluminum that is poured into a mold and then cooled. Cast piston molds are fairly expensive but are extremely durable, and once they are finished the cost per piston is very low. Cast pistons are, thus, considerably less expensive but have limits to their ultimate durability under the stress inherent in a high-performance application. When married to sophisticated electronic engine management, which prevents detonation, a cast piston can handle quite a bit of power.

Forged pistons are extruded into their shape under extreme pressure. They can be formed either in a mechanical or a hydraulic press. The forging process delivers a part with greater ductility (the ability to "bend" or deform) and material density than cast parts. Forging tools are quite expensive, wear out in service, and the process is slow compared to casting. All these factors mean that forged pistons cost more money than a similar cast part.

A cast piston and a forged piston may well have the same compressive and tensile strength in testing. The advantage of forging lies in ductility and fracture resistance. It's not so much the ultimate strength as it is the mode of failure once the limits have been exceeded. Forgings tend to go into plastic deformation, while castings tend to fracture when overloaded. This forgiving nature allows forged pistons to survive extreme power levels and the occasionally marginal tune ups. Engines beyond 500 hp are almost always going to have forged pistons.

Piston Alloys–Hypereutectic, 2618, 4032 and More

Cast and forged pistons can be manufactured from numerous aluminum alloys. Alloys are a separate subject from the manufacturing process. Some aluminum alloys may be either cast or forged, although in most cases the alloy is optimized for the intended purpose.

Many original-equipment cast pistons and most performance aftermarket ones are cast from hypereutectic alloy. Hypereutectic aluminum by definition has a high percentage of silicon dissolved within the alloy. While common casting alloys can only hold roughly 11 percent silicon in a dissolved state, the hypereutectic alloys contain up to 16 percent. The excess is dispersed throughout the piston in the form of "nodules," which are microscopically small bits of silicon. Silicon in a piston increases wear resistance, improving both the durability and strength of skirts and ring grooves. The downside of the additional silicon is that each nodule becomes a stress point, and hypereutectic pistons tend to be less fracture resistant than those made from alternate materials.

Forged pistons also have alloy choices. The two most common ones are designated as 4032 or 2618 aluminum, although many others are possible. The 2618 is stronger at high temperatures, and has greater ductility. The 4032 contains about 11 percent silicon, and is the better choice in street- or high-use race applications in which long-term wear resistance is a factor. A piston formed from 2618 alloy, which has essentially no silicon, is the better choice for racing applications where high-temperature strength and ductility are more important. While the differences between the two alloys are significant, there are numerous other design factors that impact the performance of the piston—making either one a satisfactory choice in all but the most demanding race applications.

In the case of our target engine (a high-performance FE), a forged piston is preferable because of availability and durability of these pistons. For drag racing and limited street use, the 2618 alloy is a better option. Since this engine will not go 50,000 miles before service, durability is less important than the ability to handle high loads. Forged pistons, in most applications, can handle more than 2,000 hp.

Contrary to popular information, the type of alloy does not determine the piston's strength. Forged and cast hypereutectic pistons can have identical tensile and compressive strength, while the alloy's composition determines its ability to handle high temperatures. The difference is ductility and post-failure behavior. Also, the construction method, whether cast or forged, does not determine the expansion rate. In turn, forged pistons are not always installed in the block with looser tolerances because the skirt design primarily dictates the piston-to-cylinder-wall specification.

Piston Design–Bore Diameter and Compression Distance

After choosing a material for the pistons, you need to define the best design for a particular application. In order to do so, you need to identify several key features of the engine. First is the bore diameter, followed by stroke and rod length. The latter two help to select the compression distance.

Determining the bore diameter seems straightforward, and it is. If you are building a 390-based engine, the basic bore was originally 4.050 inches. On a 428, the bore started out as 4.130 inches. The 427 had an original bore of a somewhat unusual 4.233-inch dimension. When reconditioning the bores in an engine, the common terms are something akin to ".030 over," meaning that the finished size is thirty thousandths of an inch larger in diameter than the original dimension. Most shops go larger in .010-inch increments on subsequent rebuilds, so a minimal amount of material comes out of the block with each overbore. It's relatively common to take the 427 to a more popular 4.250-inch bore first, rather than sticking to the .030-over mentality. It is worth noting that the clearance needed for piston skirts is manufactured into the piston—the bore is always assumed to be the nominal specification.

Compression distance refers to the dimension between the piston pin centerline and the deck surface of the piston. This is measured to the piston's flat top—not the top of the dome, if any. The desired compression distance can be calculated by subtracting the crankshaft stroke divided by two, the center-to-center connecting-rod length, and the desired deck clearance from the block's deck height as measured from the main bearing centerline to the cylinder head mounting surface. It is usually best to allow for a small amount of deck clearance even if your desired target is zero, to accommodate any dimensional variances.

Compression Distance Calculation Example

Block Deck Height
10.170 inches
Subtract:
Stroke divided by 2
(4.250/2 = 2.125)
– 2.125
Rod Length, center-to-center
– 6.700
Desired Deck Clearance
– .005
Desired Compression Distance
1.340 inches

Skirt Design and Oil Return

Piston-to-bore clearance is primarily dependent on the skirt profile combined with the oil-return configuration. The differences between various piston alloys in terms of growth are modest. The differences in growth between pistons made in a casting or forging process are even smaller. A 2618 alloy piston needs more clearance, and not so much because of growth—it is because the lack of silicon in the alloy makes it less tolerant of bore contact.

Oil-return design has a big effect on needed skirt clearance. A race-style design uses a series of drilled holes into the back of the oil ring groove to permit return of scraped-off oil to the pan. A passenger car piston uses a slotted oil-return passage for the same purpose. The drilled holes make for a stronger and less flexible piston skirt that needs more bore clearance. The slots make for a more flexible, but weaker, skirt. A flexible skirt can run tighter clearances and makes for a quieter engine when cold—good for your daily driver.

After the engine is up to temperature, there is no advantage in ring sealing from having a tight cold clearance. The same is true for noise. No matter what the alloy or manufacturing process, operating clearances, with the piston at normal temperatures, are going to be very similar.

Also most 2618 alloy pistons are designed for racing applications where noisy cold operation is less important than strength and friction reduction.

Dome Design, Compression Ratio Calculation and Selection

The next piece of the piston puzzle is to define the dome design. In order to do this, you need to choose the desired compression ratio and decide upon the cylinder-head chamber volume. All these things are interrelated and can have a dramatic impact on the engine's performance. You need to do careful research when selecting engine parts because pistons and heads should be complementary, and at least compatible.

Compression ratio is defined as: The comparison between the amount of air volume above the piston when it is at the bottom of its stroke, to the amount of volume when it is at the top of its stroke. This

This FE domed piston features a ceramic coating on top, which withstands greater combustion temperatures, so the tuner can run air/fuel mixture leaner while maintaining reliability. The skirt is moly coated for increased lubrication and reduced friction and enhanced scuff resistance.

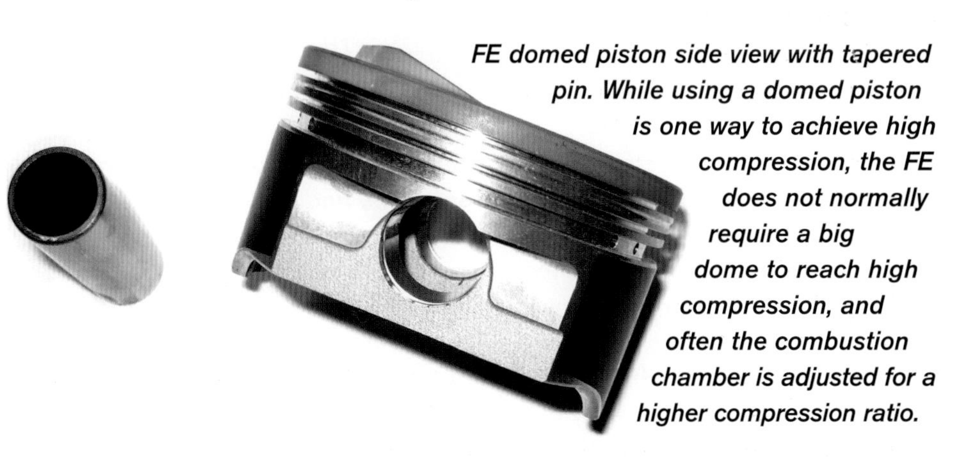

FE domed piston side view with tapered pin. While using a domed piston is one way to achieve high compression, the FE does not normally require a big dome to reach high compression, and often the combustion chamber is adjusted for a higher compression ratio.

ratio is a function of mechanical characteristics and does not change with RPM or fuel, or anything else.

To calculate compression ratio, add up all the volumes above the piston at each end of its travel. These calculations are commonly done in cubic centimeters (cc) so some conversion from inch measurements is required. One cubic inch equals 16.387064 cubic centimeters.

The volume areas to be concerned with are:

- Combustion-chamber volume of the cylinder head
- Effective dome volume of the piston, which can be "negative" if domed
- Crevice volume: the area above the piston rings but below the piston's deck
- Volume of the head gasket's cylinder opening
- Volume of the deck clearance above the piston at top center

The above items are referred to as "unswept volume" because the movement of the piston does not affect them.

The "swept volume" of the cylinder is defined as the cylinder's volume as calculated by its bore diameter and the stroke.

This formulaic approach compares the cumulative volumes of all the above items to the cumulative volume of only the first five. Volume calculations for the cylindrical items can be done using the standard "radius squared x pi x depth (or stroke)" formula, but the other values need to be either actually measured or taken from manufacturer's data (depending on your degree of faith in their published material).

Compression Ratio = (unswept volume + swept volume) ÷ unswept volume

The calculation can be done by hand, but these days it is commonly handled with a simple computer program. This allows you to quickly change the variables and find the combination that best meets your needs.

With a spherical dish race piston, the dish is deeper in the center. Also note the cylinder-location-specific asymmetrical valve pockets. The spherical dish is current technology for many OE and professional racing applications. It is said to focus combustion pressure directly over the center of the rod.

This race piston has gas ports for enhanced ring sealing. These are the vertically drilled holes in the dome of the piston. It provides a proven power gain when used with light-tension rings. These are designed for racing because in street use they can carbon up. Once the ports get clogged they're no longer effective.

Gas ports allow combustion pressure to quickly build behind the compression ring and seal it against the cylinder wall. As a result, the ports reduce ring flutter in the bore, and the powerband can be moved up in the rev range. Piston side view shows the return holes in the oil-ring groove, and a pressure-accumulator groove between the top and second ring.

An FE blower piston with a 36-cc-deep dish delivers reduced compression. Typical supercharged and turbo-charged FE engines run a 9:1 or lower compression ratio in order to get the greatest benefit from the boost available.

Below is an example using a 4.250-inch bore and 4.250-inch stroke, and common volumes for FE heads, gaskets, and a dished piston:

Combustion chamber volume
72.0 cc
Effective piston dome volume
17.0 cc
Crevice volume
1.0 cc
Head gasket volume
10.2 cc
Deck clearance volume
1.0 cc

Total unswept volume
101.2 cc

Swept volume of cylinder
$(4.25/2)^2$ x 3.14 x 16.387064 =
988.4 cc
Remember to do the cc conversion!

Using the formula, the compression ratio is 101.2 + 988.4 = 1,089.6

Compression ratio for this example is:
10.77:1

After calculating the compression ratio, what can you do with it? Raising compression is considered a guaranteed way to add horsepower. But that statement only holds true within certain constraints. A commonly referenced general rule is that you get about a 4-percent power increase per point increase in compression. This translates to perhaps 25 hp in a 600-hp engine by going from 10:1 to 11:1 in compression. This is by no means a definite value, and is not linear; the increase becomes progressively smaller as the compression ratio increases.

A number of factors determine the amount of compression for a given application. These factors include the quality of fuel, design of the combustion chamber and piston dome, weight and type of vehicle, and power adders if any—such as nitrous, blower, or turbocharger. It is far better to target a relevant range for compression, rather than fixating on one particular number.

Using pump premium gas for naturally aspirated applications is typically best when compression is set between 9.5:1 and 11:1. A heavy vehicle, such as a Galaxie with an automatic and highway gears, should be at the lower end of that range. A lightweight Cobra with a stick and steep gearing would be fine at the upper end of the spectrum. A modest sacrifice in power is well worth the ease, reduced expense, and convenience of running pump gas for a street car.

Looking underneath this race piston, you can see the forced pin oilers, which provide enhanced oil flow to the wrist pin under demanding racing conditions. The piston has also been CNC machined to make it as light as possible while retaining strength. Lightweight components allow the engine to reach a higher RPM level with the given rods and crankshaft, and RPM equals horsepower.

If you are going racing, the game changes. Power is the goal and race fuel is a necessary part of the equation. Drag racers normally find the best compression considerably higher than 12:1—often exceeding 14:1 on serious engines. Oval-track and road-race applications normally hold to somewhere between 12:1 and 13:1, in an effort to retain greater durability and handle the higher temperatures they encounter.

It is currently popular to use the calculated actual compression ratio along with the cam event data to generate a "dynamic compression ratio" (DCR) using any number of available calculators and a computer. These dynamic compression ratio numbers are sometimes useful for comparison of different build strategies. But this ratio is most valuable as a starting point, and not definitive by any means. The numbers do not take into account the various combustion chamber shapes, piston dome shapes, spark plug locations, intake configurations, or a host of other variables that enter into fuel tolerance. Use a DCR calculator to get close, but don't become fixated on any particular number; you can be misled.

Piston Top Shape and Design

Current trends for piston top design are headed toward minimizing any dome, in favor of either a flat top or a dish. After decades of research, most OEM pistons have evolved to this style as they seek improved efficiency. I cannot find any reason not to follow this trend—"stealing" a good idea *is* a good idea in this case.

Dished head pistons reduce compression when compared to a flat-top piston and symmetrical valve pockets allow use on all cylinders. The dished head is usually required for pump-gas compatibility on stroker engines, but the flat-top design works well with factory displacement combinations.

The FE is a good engine from the perspective of piston-dome design. With small cylinder-head chambers, you have the ability to get fairly high compression from a flat top or very modest dome on the piston. The earlier example showed a dish piston that achieved a good pump-gas compression ratio. If you chose to run a flat top in that particular application, you would be just under 12:1. This is the ragged edge for a street car and would perhaps need a bit of race fuel added to the tank.

Given the fact that a flat piston weighs less, and has been demonstrated to be more efficient in terms of combustion flame travel, it has become common for race builders to mill the heads, rather than add a dome when seeking more compression. For those not wanting to cut the heads, a modest dome of around .100 inch gets compression ratios approaching 13:1 with the more popular aftermarket head combinations.

A heavy piston does not offer any advantage once you reach an adequate thickness for strength. A lighter piston is easier on rods, bearings, and crank; and a lightweight package revs more freely. But the main advantage of a lighter-weight piston comes from its durability. Because load on the bottom end goes up with RPM, it is intuitive that a lighter assembly goes to a higher RPM before those loads reach or exceed the limits of the components. In a race engine, RPM equals horsepower—more "bites" of air and fuel per minute.

Other Components

Piston pins are important items that do not get enough attention in engine building. If the pin is not strong enough, it will flex, causing wear and possible failure of both rods and pistons. It is common for an inadequate pin to be the true source of issues that are blamed on the pistons.

The four ways to make a pin stronger are to increase its diameter, increase its wall thickness, reduce its length, or upgrade the materials. Lighter components allow the engine to rev more easily, but weight-fixated builders often go the wrong way on pins as a result, sacrificing the necessary strength and rigidity. Piston companies offer multiple pin designs, and are a good resource for proper selection. A larger-diameter, shorter, thin-wall premium-alloy pin is rarely a bad investment. Compared to the common 4340 alloy, a drag-race piston pin made from H13 tool steel can take advantage of the better material, using a thinner cross section to reduce weight without sacrificing strength.

Pin retainers are another subject worth touching on. The FE is blessed with floating pins from the factory. The original parts and some basic aftermarket pistons used a single Tru-Arc–style retaining clip. Higher-quality pistons went to a double Tru-Arc many years ago to prevent the clip from breaking and the pin from walking out of the piston. In the past 15 years, the Spiro-Loc clip has become the retainer of choice, often in double-clip designs for the insurance value.

Recently the round wire retainer has returned to the domestic performance market on both very-high-end pistons as well as some basic cast parts. The round wire better distributes pin side loads across the entire pin boss, rather than focusing them on the edge of the snap-ring groove. But pin-lock failures with Spiro-Locs are scarce, and the important thing is to be certain that you are using the correct lock-clip design for the pistons you have, as well as matched piston pins. Piston pins used along with round wire locks require a large bevel/chamfer to work properly.

Pin oiling is normally of the "forced" variety on race pistons, with a feed passage from the oil-ring groove into the pin boss. While considered the norm in the race world, the forced design likely has little, if any, real advantage other than a limited-lube environment with a dry-sump or vacuum pump. Every OEM piston, along with many aftermarket parts, uses a broached groove for oiling the pin. And failures in those pistons from inadequate lube are very rare.

Piston-Ring Grooves and Piston Rings

The ring grooves on the piston are really a part of the piston-ring package. They just happen to be incorporated into the piston. The ring grooves are a sealing surface that is as important as the rings and the cylinder wall.

Ring grooves need to be extremely flat, non-wavy, and perpendicular to the bore in order to function correctly. A well-designed ring groove has a small radius at its root (where the back is) for strength, and has a small degree of vertical uptilt. Vertical uptilt is cutting the ring groove at a slightly upward angle to compensate for changes in the piston shape due to temperature. The piston always runs hotter on top and grows more there as a result. None of these features are readily visible to the naked eye, and all are extremely difficult to measure. You need to trust your piston supplier, and assume that there is a reason that cheaper parts are, well, cheaper. Thermal expansion rates are alloy dependent and are the same. But the better piston compensates for the dimensional change through design and machining, so that ring seal is not compromised as temperatures rise.

Piston Rings

A piston ring is a sliding dynamic sealing device. It functions and seals through combustion pressure that forces it against the cylinder wall. The top ring does nearly all of the real combustion sealing work. The second ring is actually an oil control item and has a tapered face, which acts like a squeegee to develop and maintain a thin film of lubricating oil. The third ring is entirely devoted to oil control, scraping the majority of oil off the cylinder walls and sending it back down to the pan. Each ring in the "stack" can be optimized to do the best possible job.

Piston-ring dimension specifications can be confusing; ring "width" is considered to be the distance from the top to the bottom of each ring. For example, a common ring width is 1/16 inch. Radial-wall thickness is the dimension measured from the bore contact surface to the inside diameter. Radial-wall thickness dimensions are often specified to an SAE "D-wall" standard of bore/22. Thus a piston ring for a 4.250-bore engine would have a standard radial-wall thickness of .193 inch. Current trends are for ring cross sections to be much smaller than in the past, with widths of .043 inch and smaller and non-standard radial-wall thicknesses of .160 inch or less.

Ring tension is measured as the average pressure exerted against the cylinder wall by each ring. Old theory held that higher-tension rings sealed and controlled oil better by forcing the rings against the cylinder bore. Current practice uses a reduced cross section (thinner in both width and wall thickness), which allows the rings to be flexible and conform to the bore—enhancing sealing while reducing friction. Friction reduction and sealing improvements result in more power.

Piston-Ring Selection

The top ring in a performance build is made from either steel or ductile iron. Inexpensive cast rings fracture when subjected to severe operation loads and should be avoided in all but the most stock rebuilds. The top ring has a plasma-moly coating on a barrel-shaped cylinder contact face. The width of piston rings is measured "top to bottom." The most popular dimension is 1/16 inch, and should suffice for the vast majority of builds. Reducing this dimension delivers improved sealing due to greater bore conformability and reduced ring weight. Going to a ring thickness of .043

The typical piston-ring set used in all modern engines (including the FE) has a combustion sealing ring up top, a scraper in the second position, and a multi-piece oil-control ring in the lower groove. Piston rings have to seal against both the cylinder wall and the ring groove for efficient combustion. Reducing friction through low tension and smaller cross-section parts improves performance. The challenge lies in getting the power without sacrificing oil control in the process.

The top ring normally has a gray contact surface, and often has a red paint stripe. A dot or pip mark indicates the upper surface. If not clearly marked, always install with the inner bevel up.

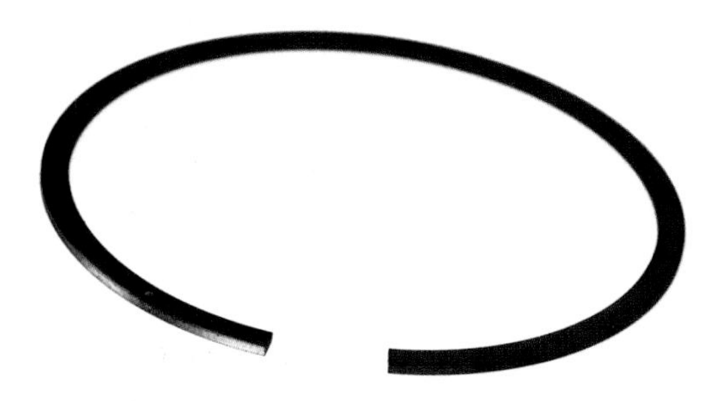

Second rings are usually tapered and have a black phosphate surface. This one shows a "pip" mark on the upper side. The inner bevel always goes down on a second ring. The second ring is primarily an oil-control device—really a "squeegee" serving to maintain a very thin layer of oil on the cylinder wall.

inch or less is a good strategy in a race application, but has only modest benefits in a performance street car, considering availability and expense.

The second ring is almost always cast iron with no face coating. As referenced earlier, it is an oil-control device. The second ring also benefits from reduction in cross section—both in width and radial thickness as measured from the inside diameter to the outside diameter of the ring. A thinner, smaller second ring does a better job with reduced drag (more on that below).

Oil rings are comprised of three pieces: an expander and two rails. The expander is normally a stainless-steel band that has been perforated, folded, and formed into a circle. It is designed to spring load the rails against the cylinder wall as well as against the oil ring groove of the piston. The rails are simple chrome-plated steel rings that contact the cylinder walls and ring grooves. As an assembly, these parts clear all excess oil from the cylinder walls on each stroke. The tension of the spring expander determines how "hard" the contact is, and how much oil is cleared or left for the second ring to work with.

As with the other rings, the current trend on oil rings is for reduced radial thickness, smaller widths, and reduced tensions.

The ring summary goes as follows:

- On a race engine, I would go straight to a .043-, .043-inch, 3.0-mm ring package with a reduced radial wall thickness.
- On a street engine, I would stick to the traditional 1/16-, 1/16-, 3/16-inch package.
- On an "in between" deal, I might consider the 1/16-inch top, a reduced radial wall thickness 1/16-inch second (often referred to as a "back cut" ring), and a low-tension 3/16-inch oil ring. This one might use a touch of oil at part throttle, but should still be tolerable on the street and run hard at the strip.

Piston rings make up a large percentage of engine friction, and are equally important for producing horsepower through adequate sealing during combustion. Reducing friction with low tension and smaller cross section parts yields measurable benefits, as long as you don't go too far and sacrifice oil control. Oil in the chamber causes detonation and subsequent damage—you can go too far. In addition, oil smoke out the exhaust is never a good thing on a street engine. It's the risk-versus-reward scenario; it's best to err on the side of caution if you are not willing to find or finance finding the "edge."

Piston-Ring End Gaps

The quick strategy is to follow your chosen ring manufacturer's

A Speed-Pro SS50U oil ring has plastic inserts that are color coded for identification for bore size and tension. The little plastic squares prevent overlap during installation, will not go anywhere, and can remain in place. Hastings-style oil expanders don't use the plastics, but you need to be careful to avoid overlapping them during installation.

suggestions for ring gap sizing. But there is a lot more to this than may first meet the eye. The top ring gap needs to be large enough to prevent ring butting at the highest temperatures the engine will see. In most naturally aspirated engines this translates to somewhere between .003 and .004 inch of gap per inch of bore diameter. At operating temperatures this gap area becomes smaller, approaching zero in an optimized race engine. On engines running boost or nitrous, the ring end gap needs to be increased due to the heat generated, usually by at least 20 percent.

Second rings used to be thought of in the same fashion. Because they run at lower temperatures the gap recommendations were smaller than those for the top. A change in theory, followed in practice by most OE manufacturers, has been to move to a second ring gap larger than the one for the top. This is done to relieve any accumulation of pressure between the top and second ring, and in turn allows the top ring to remain seated against the piston's ring groove for better sealing. Recommendations are for second-ring gaps to be around .0050 to .0055 inch per inch of bore diameter.

Oil ring rails do not see much heat, and usually only need to be inspected for adequate gap. Roughly .010 to .012-inch total gap is sufficient.

Ring filing is best done with a rotary tool for the purpose. Filing only one side of the gap allows you to use the uncut side as a reference to ensure that you have the gap surface straight. Once you have the proper size, use a knife-sharpening stone to edge break any sharp corners left from the filing. You do not need or want a chamfer—just a clean, square edge with no burrs.

Alternative Ring Concepts—Gapless, Napier and Dykes

There are several specialty rings available in the performance market. Some of them make claims of greater performance, durability, oil control, or all three. Conventional piston-ring designs have been under continuous development at the OE level for decades in pursuit of more power, efficiency, and reduced emissions, and they are fairly well optimized.

Others may differ in their opinions, but I am not a supporter of the gapless-ring theory. At the time of this writing there is not a single OE engine using a gapless-style ring, despite claims of tremendous benefits. It does not matter whether the engine is in a quarter-million-dollar exotic or a max-effort fuel-economy vehicle. Despite huge research budgets and intense competitive and government pressure, they all run a conventional-style piston-ring package. An engine like the one found in a Corvette Z06 or an NSX Honda can justify titanium connecting rods, so it is not a cost-driven decision. There is a lesson in that data for those caring to listen.

Generally speaking, if a piston ring claims more than an incremental gain in any performance metric it is likely to be too good to be true. If the claims read like something out of a nitrous catalog it's virtually guaranteed to be marketing hype.

The Napier-style second ring is a different story. The Napier ring has an undercut around the outer diameter, which reduces the effective cross section of the piston ring while retaining the material thickness that lies within the piston's ring groove. It is used in numerous OE applications and has comparable friction-reduction benefits in a performance engine.

A Dykes-style ring has an inverted "L"-shaped cross section with the cutout part going into the piston ring groove. This again reduces the ring's cross section and allows combustion pressure to better reach the backside of the piston ring, thus enhancing sealing. This ring design is somewhat out of favor these days, but is still well accepted in race applications where gas-ported pistons are not permitted, along with use in supercharged race engines.

Leakdown testing is touted as a way to gauge ring sealing, but in practice has major limitations. Leakdown tests are performed with 100 psi of shop air, at room temperature, on a non-moving piston. There is no way that they serve to simulate the performance of rings that operate at temperatures of 700 degrees F, seal nearly 2,000 psi of combustion pressure, and are traveling over 4,500 feet every minute.

Your best bet functionally and financially is a conventional ring set with reduced radial wall thicknesses and ring tensions appropriate to the use of the vehicle.

CHAPTER 5

THE OILING SYSTEM

A lot has been written about the FE oiling system over the years. Many have suggested that the factory system is inherently flawed, and that major modifications are warranted. Having built numerous FE engines over the past 30 years, I disagree. The FE factory oiling-system strategy is essentially identical to that of most other Ford V-8s, and requires only modest detail work and attention to assembly in order to function perfectly well. Dramatic alterations are unnecessary in the vast majority of applications.

A typical Melling high-volume FE oil pump is a gerotor design; a 1/4-inch hex shaft drives it. High-volume pumps typically deliver 20 to 25 percent more oil volume than common replacement oil pumps. They're not always a necessity, but will raise the oil-pressure reading. Low-idle oil pressure is a common concern with the FE, although it rarely causes any problem with the engine. Melling offers a variety of pumps for the FE, including standard, high-volume, and high-pressure variations.

Oil Pumps

The FE engine uses a gerotor-style oil pump. It's mounted to the front left corner of the engine block inside the oil pan. The distributor drives the pump through a 1/4-inch hex-ended intermediate shaft. A pressure bypass spool valve is integral to the pump, and a selective spring controls maximum pressure. Some original pumps were aluminum, but the replacement pumps are cast iron. Replacement pumps are offered in standard volume and pressure. But several other options are available. You can choose from high-volume designs with a deeper pump body, high-pressure designs with stiffer bypass springs, and high-volume/high-pressure versions that combine the stiff bypass spring with the deeper pump body.

The bypass spring serves as the effective limiter for peak oil pressure when the engine is cold or at high RPM. The cumulative area of all bearing and valvetrain clearances in the engine defines the pressure at any time below peak, such as at idle. FE engines are noted for having fairly low-idle oil pressure, and therefore a hot idle reading of 15 to 25 pounds is common with standard-volume pumps. A higher-volume pump generates higher-pressure readings at all points below peak and serves as something of a confidence booster, along with having value in higher-output engines where clearances are intentionally larger. The oft-repeated

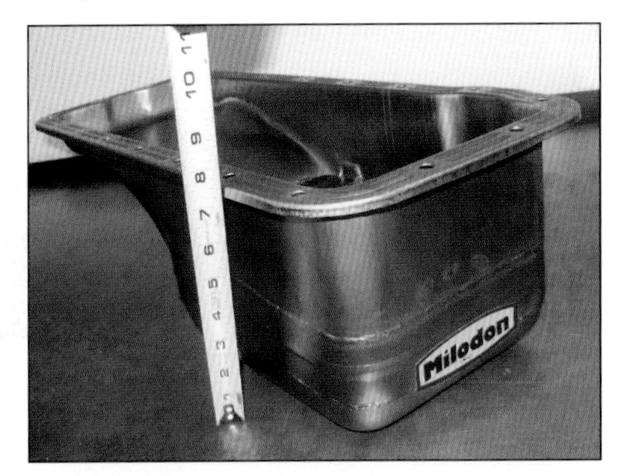

Measuring pan depth and pickup-to-pan clearance is simple on a deep-skirt engine like the FE. This is a critical step when building any engine because you need to ensure that the pump-to-pan clearance is adequate. The last thing you want to discover is oiling-system problems after initial start-up and have to deal with subsequent engine damage.

The oil-pump windage tray and pickup have been assembled here. Some need to be adjusted for proper clearance. I find it easier to mount the tray to the block first with a few bolts before assembling the pickup onto the pump. This permits easy inspection for interference.

This a factory oil-return tray in position on the head. While it's an effective part, it doesn't work with many aftermarket rocker systems. The long-fingered return trays direct oil through the opening in factory intake manifolds. Even if the tray is modified to fit your rockers, many aftermarket intakes are not designed to accommodate the tray's fingers.

The mid-sump race pan puts the oil sump farther back in the chassis, which lessens the oil control negatives associated with front sump pans in drag-race applications. It uses an extended pickup tube supported by a main cap bolt, so that proper pump depth and position is correctly set.

adage of needing 10 pounds minimum pressure per thousand RPM is still valid. Some builders desire higher pressure values as a matter of personal preference; and those can be had with the high-volume pump configurations.

Some have reported that a high-volume pump pulls the stock oil pan "dry." Therefore, the drainback speed isn't fast enough to adequately keep the oil pan filled and, as a result, delivery is more of an issue than back speed. Ford specified that factory 428 CJ engines carry an extra quart of oil, so the 5-quart pan was the same depth and volume while actually holding 6 quarts. The combination of high RPM and the front sump can apparently cause problems in certain conditions, and the high-volume pump may well exacerbate these conditions. The solution seems simple though. If you want to run a high-volume pump in an otherwise stock-style application, add an extra quart of oil.

At Survival Motorsports, we tend to use a common Melling M57HV high-volume oil pump on the majority of engines. We disassemble them, detail the inlets and outlets by removing any obvious burrs or impediments to flow, check clearances, and inspect for free movement of pump components and the bypass valve upon reassembly. Pump rotor-to-cover clearances are critical, and they need to be level and around .001 to .003 inch at the most. If the clearance is larger, bleed off causes idle oil pressure to suffer. The edges of the gerotors need to be nearly sharp for greatest pump efficiency, so they may be deburred but should not be chamfered. On our personal race engines built at Survival Motorsports, we will use a standard-volume pump, trading off the small increase in power against the reduced oil pressure, but do not typically do this on a customer engine.

Precision Oil Pumps offers several blueprinted and safety-wired oil-pump options if a ready-to-install "out of the box" package is desired. This company also offers an extra-high-volume pump, which has merit in aluminum block and racing applications.

We commonly install oil pumps with high-quality fasteners and a bit of Loc-Tite. We safety wire the bolts for those applications that are going to see extended running and high vibration, such as road-race cars or off-road trucks. Ford pumps have a thin paper gasket between the block and the pump. We use a *very* small amount of sealer on the gasket.

Pump Driveshafts

The factory FE pump shaft is a simple piece of 1/4-inch hex stock that connects the distributor to the pump center rotor. A serrated washer is pressed onto the pump shaft to prevent it from coming upward and loose during distributor removal. Some FE truck engines use a larger 5/16-inch hex drive, but these drivers are not typically used for high-performance engine builds.

The ARP FE oil-pump drive is a stronger replacement for the small-diameter factory piece. Made from heat-treated, high-grade chromoly steel, the ARP design features a wider-diameter, heavy-duty shaft for racing and high-performance use. ARP shafts withstand the added torque of high-volume and high-pressure oil pumps as well as heavy-viscosity lubricants.

The factory shafts had a reputation for twisting into a "barber pole" when debris jammed into the pump, causing an abrupt loss of oil pressure and subsequent engine failure. Often, pieces of valvestem seal or timing gear teeth were ingested into the oil pump on high-mileage engines, which caused the oil pump to jam and the engine to fail.

For longevity and reliability, we use an ARP heavy-duty pump driveshaft in all of our builds at Survival Motorsports. This upgrade isn't absolutely necessary because shafts only fail when debris stops the pump, and debris should never be present in a performance engine, but it is inexpensive insurance. The heavy-duty shafts have a large-diameter full-round center section with the drive hex formed at each end. Precision Oil Pumps offers shafts similar to the ARP parts, along with a 5/16-inch pump drive end option to match its specialty pump line.

Oil Pump Pickup Screens

The pump pickup is a matched item to your choice of oil pan. Most FE pump screens mount to the side of the pump with a pair of 5/16-inch fasteners. FE automotive oil pans are front-sump types, thus the pickup is pretty short and fairly vertical. Some trucks use a center/rear sump pan with a long pickup tube. Both work perfectly fine, but the long tubes take more time to prime during engine prelubing, causing a couple tense moments waiting for the gauge to move. Higher-end drag-race oil pans may also have a center sump design. The factory-style front sump is not well suited for maximum acceleration in which the oil is forced toward the rear of the engine. The center sump can be a challenge to fit into the car, though, often requiring cross-member and/or steering modification.

I do have a preference to the mesh-screen design as used by Moroso over the perforated metal seen in some others. Every OEM oil pump pickup I've ever seen used a mesh screen. The screen should be between 3/8 inch and 1/2 inch off the bottom of the oil pan. Factory screens use a folded sheetmetal strap to prevent the screen from bottoming out against the pan, which is a useful feature that most aftermarket performance screens do not have. Factory screens also have a provision to allow oil flow even when the screen becomes obstructed. Race parts do not include this feature, nor should they require it.

Use a very, very small amount of sealer on oil-pump screen gaskets to prevent any possibility of air leakage on the inlet side. This is very important because the inlet side of a pump is very sensitive to air leakage. Think of having a pinhole in a drinking straw. Also use red Loc-tite on the mounting fasteners. This makes removal more difficult, but you need the fasteners to be secure. They can

Here are a couple of Milodon oil-pump pickups: one for their replacement pan (right) and one for their deep pan (left). The pickups feature larger-than-stock-diameter thick-walled tubes, which are precision bent.

be safety wired in high-vibration applications. The center-sump-style screens often have a support that mounts to a center main cap bolt as a way to eliminate vibration-induced damage.

Oil Filter Mounts

The FE oil filter mounts to an aluminum casting, which is bolted to the front driver side of the engine block. The vast majority of these orient the filter in the vertical position. Some truck parts offer alternate filter positions for special needs. Within the common passenger-car mountings, there are a few variations to be aware of. Most physically function on any application, but some are tilted rearward slightly, allowing improved sway-bar clearance. The performance-style mounts have larger cavities for enhanced flow, but standard mounts can be modified for comparable results.

Cast-aluminum factory oil-filter mounts, such as this Cobra Jet one, have deeper cavities on the block side and are a better design than some. Cast- and billet-aluminum mounts are available for the entire range of FE engines.

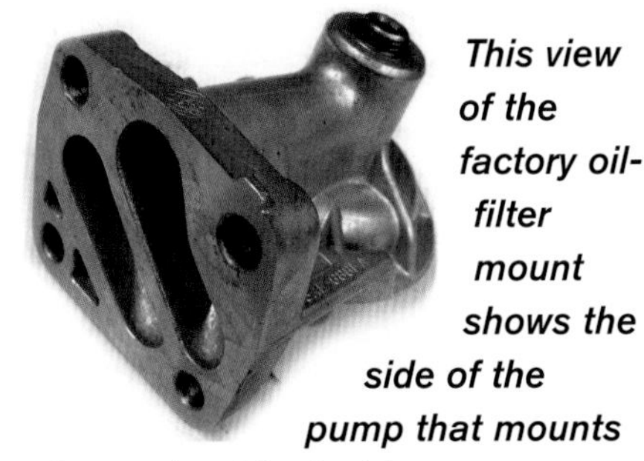

This view of the factory oil-filter mount shows the side of the pump that mounts to the engine-block side.

Oil-filter-mount gaskets usually come with only a couple of punched round holes, instead of having the full opening matched to the mount.

This filter mount is shown with a modified (bottom) and an unmodified (top) gasket. Opening the gasket holes to match the bean-shaped slots in the filter mount lessens the potential for leakage, and from having bits of gasket paper entering the oiling system.

The factory oil-filter adapter is installed on the engine, simply bolts to the side of the engine, and the dipstick goes right behind the mount. Some truck engines also have a compressor drain opening in the block near the dipstick, which must be plugged for passenger-car use.

This is an oil-cooler adapter for 428 SCJ engines. It allows you to install an oil cooler and maintain lower oil temperatures in racing and extreme high-performance use. Blue Thunder now reproduces these.

A typical Melling high-volume FE oil pump is a gerotor design; a 1/4-inch hex drives it. High-volume pumps typically deliver 20 to 25 percent more oil volume than common replacement oil pumps. They're not always a necessity, but will raise the oil-pressure reading. Low-idle oil pressure is a common concern with the FE, although it rarely causes any problem with the engine. Melling offers a variety of pumps for the FE, including standard, high-volume, and high-pressure variations.

I modify the mounting gaskets for the filter mounts. The normal gasket has a pair of holes that match the block. You can take a handicraft knife and open up the holes to match the bean-shaped openings on the filter mount, which eliminates the chance for loose gasket material to impede oil flow.

Precision Oil Pumps offers a billet-style mount for those seeking that appearance. Also available are a variety of remote filter and cooler adapters for road-race applications.

Block Oiling Circuitry

With the oil pump and filter hardware defined, you can move on to the block itself. While hard to describe, a few minutes spent probing through passages with a piece of welding rod or coat hanger makes it easy to understand the oiling strategy on an FE block.

On a typical center-oiling-style FE, the oil leaves the pump at the block mounting surface. It makes an abrupt 90-degree turn and heads out toward the filter. The pump outlet is an oblong, roughly 1/2-inch-diameter opening, while most factory blocks have a straight 5/16- or 3/8-inch drilled hole for an inlet. Use a carbide burr on a die grinder to open that passage up to match the pump outlet. Then use a 60-grit cartridge roll (aka, "tootsie roll") on the die grinder to blend and contour the turn, so it has a smooth transition.

Contrary to popular belief, the oiling system on an FE engine is not flawed or subpar. This illustrates the path of the oil from the oil pump to the filter mount. Be sure that the passage is free of any obstructions or machining flaws. With any max-performance engine, you need to maintain optimal oil-flow volume.

That same cartridge roll will be used at the filter-mount openings to clean up any machining steps or burrs.

Oil routes from the side of the block and through the filter. It then returns into the block through a passage that is angled upward toward the passenger side, and then it lubes the front main and cam journal on its way through. This meets up with another passage just forward of the passenger-side lifter bank. This passage runs back and upward toward the driver side, intersecting the front-to-back center galley and ending with a plug right alongside the distributor hole.

This shows the side-oiler passage and oil route on a 427 from the filter mount into the block. As the name implies, the side-oiler passage runs along the driver's side of the block, routing oil to main bearings for the purpose-built 427 race engine.

You do not need to significantly alter the oiling system on an FE unless it will be extensively modified. The stock system provides adequate oil volume and pressure for most combinations up to 500 hp. This photo shows the oil path through the side-oiler passage, which feeds the main bearings.

The main center passage in turn carries oil to vertically drilled passages and feeds each cam bearing along the block. There are annular grooves around each cam bearing in order to carry oil to the vertical passages feeding the main bearings. The feeds to the main bearings are vertically in line with the respective cam journals and are often slightly offset from the feed holes in the main-bearing inserts. It is a common practice to use a die grinder with a small burr to blend the block feeds so that they match the bearings. If you choose to make this modification, remove only the smallest amount needed and go only a quarter inch into the opening because this is a known crack-prone area in non-cross-bolted blocks.

FE block galleys all end with plugs, either press-in or 1/4-inch NPT screw-in. I prefer the screw-in design and convert to screw-ins on press-in-type blocks if the casting is thick enough. When I cannot add screw-in plugs, I use epoxy as a plug

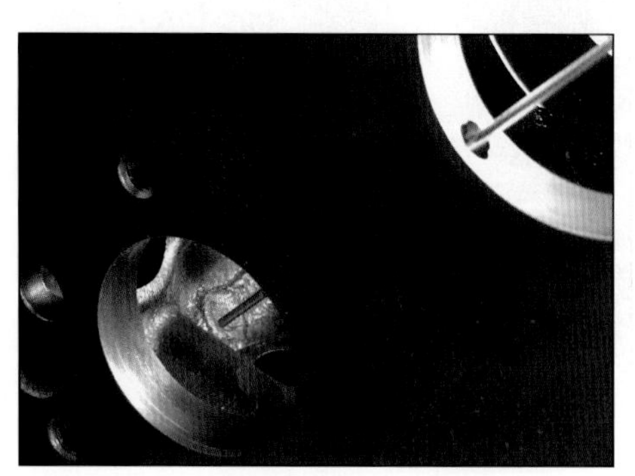

Most FE engines have a top-oiler engine oiling system, and therefore the oil path routes from the top end down to the crank. The top oiler's central main oil galley runs front to back through the lifter valley. It feeds oil to the cam tunnel and then down to the main bearings.

Factory rocker drain tins were offered in a variety of styles over the years. They do a good job of directing drainback oil from the heads, but often do not fit many aftermarket rocker systems or intakes.

sealant, and stake the press-in plug with a few light punch marks around its circumference.

The center galley also feeds the lifters, in most cases, through a pair of angled galleys found toward the rear of the block's lifter valley. On the typical FE, the lifter galleys feed the lifters only, which is different than the common Ford engine. Factory solid-lifter engines had no oil to the lifters and relied on splash; the lifter feeds were undrilled. It used to be common practice for race-engine builders to block these feed passages on solid-lifter applications, but now the trend is to either leave them open or restrict them using a drilled set screw. The passages are easy to tap to 3/8-inch thread size. When they are restricted, I use a .060-inch drill, preferring to keep some oil flowing to the lifter bodies.

A pair of galleys (that run front to back) feed the lifters and have plugs at the rear of the block. The cam retainer plate bolt plugs the front of the passenger-side galley, while the driver side has a plug that hides behind the distributor hole.

On FE engines, the rocker-arm system feeds lubricant through a series of convoluted passages. They start with passages leading from the number-2 and -4 cam bearings upward to the cylinder-head deck. Once there, they go to an opening in the head, around a head bolt, and up to a feed hole alongside one of the center rocker mounting bolts. On factory installations, the rocker bolt in that position has a reduced shank diameter to improve oil flow. Oil travels around that bolt, through the pedestal, into the rocker shaft, and then out to each rocker arm.

Side-oiler 427 engines are different. The upward-angled passage coming out of the filter first intersects a front-to-back passage that runs alongside the block on the driver side. This serves as the primary feed for oiling to the crankshaft, and therefore feeds the mains and rods before any other parts. This routing strategy is also found in new race engines and is called "priority oiling." These blocks generally do not support hydraulic lifters, and lack the annular oil grooves behind the cam bearings. They require a special cam- bearing set with extra passages, and a camshaft with grooves in the number-2 and -4 bearing journals in order to supply oil to the rocker arms.

This shows the oil-feed path from the block's deck to the cylinder heads. The bottom of this passage intersects the annular groove in the cam tunnel. The oil route to the rockers on an FE is somewhat tortuous. Once the oil travels through the head gasket, it runs a short distance across the face of the head and then up alongside the nearest head-bolt hole.

Normal side-oiler cam tunnels have no annular grooves. Cutting a partial annular groove into a side-oiler cam tunnel allows the block to be used for an SOHC conversion, and also permits the use of readily available "normal" FE cam bearings.

Often oil flow is restricted to the rocker arms, so more oil flow is maintained across the main bearings. This common FE modification can be made either in the head or at the deck. Deck restriction tapping is simple during block prep.

There are a couple variations on the above themes. Some late-production 427 engines, as well as many service blocks and most aftermarket blocks, have a side-oiler main-feed design along with hydraulic lifter capability. You may also find marine

When using a Fel-Pro gasket with a Genesis block, it's common to find that the oil-feed hole on the block doesn't align with the gasket. The fix is to either drill an extra hole in the gasket or put a shallow groove in the deck.

To restrict the oil-feed hole at the deck, the oil gallery has been drilled and an Allen-head setscrew has been installed.

engines, which are visually cast as side oilers, but these have a center-oiling strategy. Often, these engines have a portion of the side-oiling galley casting machined off for mounting clearance, and can be identified by the brass core plugs—as opposed to the common steel versions.

Pressure bypass systems are found at the rear end of many side-oiler or center-oiler blocks. Consisting of a spring and a spool valve, these are used to regulate peak oil pressure and require an extremely high-pressure bypass spring in the oil pump to be effective. It has become common practice to disable these valves with steel shims or tubing and rely on the "normal" oil-pump-mounted bypass valves.

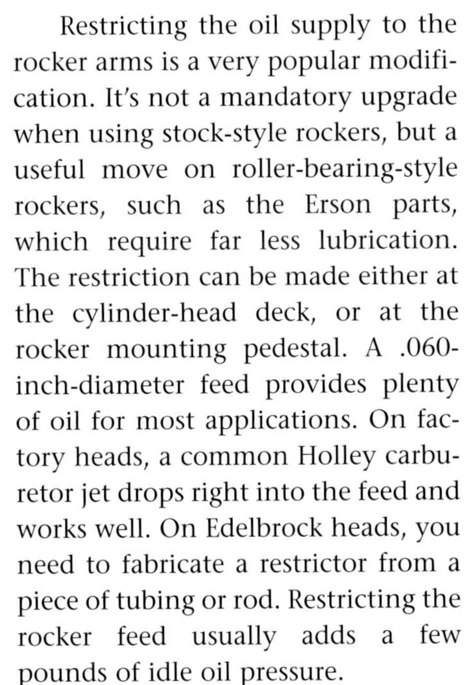

Restricting the oil supply to the rocker arms is a very popular modification. It's not a mandatory upgrade when using stock-style rockers, but a useful move on roller-bearing-style rockers, such as the Erson parts, which require far less lubrication. The restriction can be made either at the cylinder-head deck, or at the rocker mounting pedestal. A .060-inch-diameter feed provides plenty of oil for most applications. On factory heads, a common Holley carburetor jet drops right into the feed and works well. On Edelbrock heads, you need to fabricate a restrictor from a piece of tubing or rod. Restricting the rocker feed usually adds a few pounds of idle oil pressure.

Valvetrain drainback tins sandwich between the rocker pedestals and the cylinder heads, and come in a couple of styles. The earlier parts have long, finger-shaped drains that extend into the openings in the intake just inside the valve covers. The more common tins have shorter fingers that serve the same purpose. The tins are not compatible with many aftermarket rocker systems. Some intakes also do not permit the use of the drain tins. The only other drainback concern lies with the front and rear corner drains in the heads.

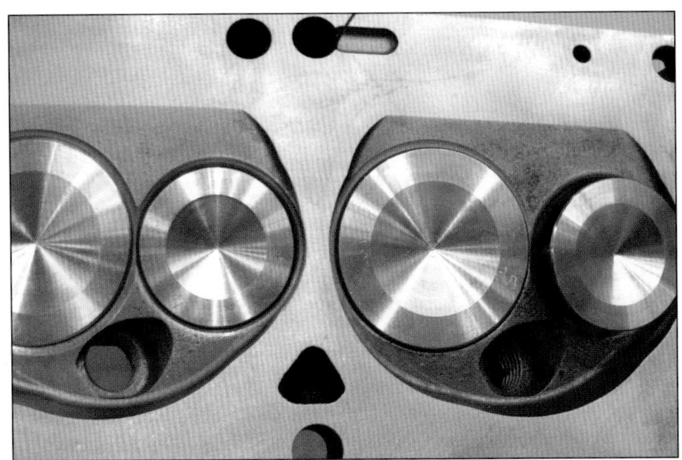

Oil-feed route is shown from the cylinder head deck to the nearby head-bolt hole. The oil flow from the block's deck intersects with the machined slot at the right-side end, travels across it, and then goes up into the bolt hole.

The silver-colored steel rod shows the oil-feed route to the valvetrain. Note the distance between the rod and the head-bolt hole. That distance coincides with the feed slot in the cylinder head.

Here is the oil route through the head to rockers in a standard FE design. However, the Blue Thunder head and T&D rocker system do not use this convoluted route. Instead, the oil is routed through the pushrods.

These drain oil around the corner head bolts. Earlier Edelbrock heads had very small drilled openings that were not in line with the openings in certain intake manifolds and gaskets, and therefore prevented proper drainback. This has been corrected in later heads, but should still be checked on assembly.

Windage Trays

Factory high-performance FE engines used a stamped-steel windage tray with a series of louvers for oil drainback. These are sandwiched between the oil pan and the engine block using two oil pan gaskets. Aftermarket windage trays are available from a number of suppliers and are either the louver-style or a screen-mesh design. In dyno testing, I have never seen any horsepower from a windage tray on an FE. The improvements in oil control under acceleration are more important in any case, which makes a windage tray a good investment. When installing a windage tray, it's important to check for crankshaft and connecting rod clearance, especially when using a stroker crank. Also be sure to test fit the dipstick; sometimes the tray needs to be trimmed for clearance so that the stick does not bend back up into the crankshaft when inserted.

Canton screen-style windage tray clears most combinations. But you still need to conduct a careful inspection to be sure it does not come in contact with any other parts. In particular, make sure that it does not come in contact with the dipstick. You may also need to add clearance for the pump pickup when using a Moroso pan.

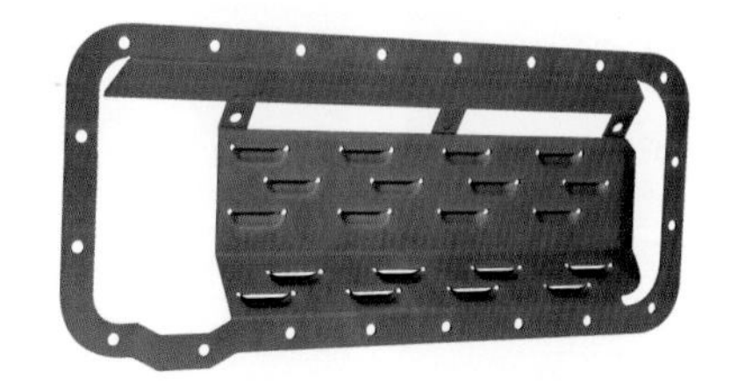

A windage tray controls oil slosh and foaming, so the oil-pump pickup keeps a steady supply of oil routing to the engine. This louvered-design Moroso tray has recently been deepened to accommodate the popular stroker crankshafts.

Oil Pans

The oil-pan mounting flange on an FE engine is flat all around, and the gasket is a single piece of cork or rubber composite without the

A Moroso T-type oil pan is a popular choice because it provides more oil capacity while delivering street-safe ground clearance. They are also offered with road-race-style baffles and trap doors for improved oil control in high-speed cornering.

separate molded rubber ends found in other engines. FE oil pans are available in numerous configurations. Most factory automotive pans have a front sump with a shallow rear section which has a couple indentations where the tie rods swing close to the engine. Some truck pans as well as specialized race pans have a center or rear sump.

Aftermarket performance automotive pans have increased oil capacity and take one of two design strategies by using a deepened sump or a standard depth with a T-style base as offered by Moroso and Canton. The "T" pans are often designed to deliver adequate ground clearance for road-race or street applications and incorporate a series of trap-door-style anti-slosh baffles intended to keep oil centered around the pickup screen.

In a wet-sump system, a very deep fabricated pan provided the only dyno-proven horsepower improvement I've documented, and this pan would be impractical in any application. The horsepower improvement was very modest. In any race or street vehicle oil control is realistically the main factor of concern.

Dry-sump pans use a series of suction sections on a remote pump to effectively vacuum oil out from below the engine. There should be a measurable power advantage to this expensive design package, which is commonly employed on professional-level road-race and drag-race applications. Although I have not had the opportunity to personally dyno test a dry-sump FE, the simple fact that every Pro Stock and NASCAR team employs the dry-sump design speaks volumes about its effectiveness.

The Milodon replacement-style oil pan is similar to the factory part in depth and design. A stock-style pan, like this, is compatible with factory chassis components and exhaust systems, as well as with most headers.

An inexpensive pressure gauge is mounted to the filter adapter for prelubrication. This allows you to verify that the engine has been adequately lubricated before initial startup. It's a wise insurance policy to protect your investment.

Milodon's deep-sump pan is essentially the same as the factory Cobra Jet part, but a 2-inch-wide band is welded into place to deepen the sump and some additional baffling is added. It's inexpensive, works well, and baffles the oil movement effectively.

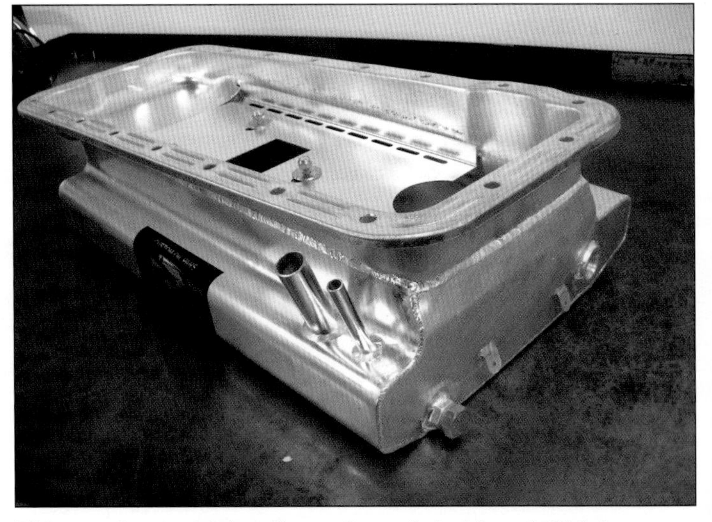

This road-race-style oil pan from Aviad is a faithful duplicate of the original Ford racing pan and remains very popular for building Cobra replicas. It includes the oil pickup and a dedicated windage tray for a reliable road-race-oriented oiling system package.

CHAPTER 6

CYLINDER HEADS

The C8OE-N Cobra Jet head's combustion chamber is shown in detail. The chamber is conventional "as cast" FE wedge material with a nominal volume of around 72 cc. Some Cobra Jet heads can be dramatically different from the published volume, so you really need to determine the volume of the chamber. This example has been slightly modified: Note the clearancing around the valves.

Ford FE cylinder head selection has undergone a complete metamorphosis over the past ten years or so. In past endeavors, one had to choose between modifying common passenger car heads or trying to find rare and expensive OEM high-performance production castings. Now there are a number of high- performance aftermarket heads for building 600- to 900-hp FE engines. Although I briefly cover some of the original high-performance head designs, I focus on readily available heads for high-performance engine builds. Also, this book is not an exhaustive document for determining the correct numbers for a given application.

True race-engine builders are either rules limited as to what they can run for castings, or they use aftermarket castings only as a starting point. They are also confident enough mechanically and financially to make radical modifications as they pursue the last possible increment of power. Therefore, these race builders start with aftermarket heads and make substantial modifications to the combustion chambers. In a similar vein, I touch on some of the current modifications and ideas for directional context and perspective, but focus on features that can be reproduced in a more-budget-friendly environment.

The average builder of a street/strip-oriented performance FE engine most likely uses either an aftermarket aluminum head or a modified factory casting with minimal concern for correct date codes or original application. My intent is to give a solid background on what's now available, what can be done at a rational cost, and what to look out for—both positive and negative.

Common Characteristics of FE Cylinder Heads

Ford FE engines use a common symmetrical cylinder-head casting for both banks of the engine. Most have two 3/8-inch threaded holes for accessory mounting on the forward, driver side of the head. All common FE heads share basic architecture that features 10 head bolts and 4 fastener points for mounting the valvetrain. With roughly one third of the valve cover located over the intake manifold, an FE head is fairly narrow and light compared to most traditional V-8 cylinder heads.

Combustion chambers are an inline "wedge" layout, with intake and exhaust valves on a common plane and a 13-degree angle relative to the deck surface. Combustion-chamber volumes range from 58 to 88 cc, the smallest being early 352 items, and the larger being factory 427 high-performance parts, neither of which are frequent finds for the casual builder. Chamber volumes on the more common heads are at 68 to 74 cc, which is fairly small compared to other large-displacement engines, but it allows high compression with flat-top pistons. Factory FE heads all use the older-style 13/16-inch hex, 18 mm spark plugs.

Valves and ports are ordered differently than most Ford engines, with siamesed intakes in the center of the head and an exhaust port at each end. The order viewed from the front of the engine is E-I-E-I-I-E-I-E ("E" = exhaust, "I" = intake).

From this angle, you see the intake face of the head. The blue Fel-Pro gasket tells you that the engine has been apart, but that should be no surprise in an engine that's 40 years old. When inspecting old heads pay attention to the relative height of the valve tips because they should all be the same. Look for broken fasteners, particularly on the exhaust flange.

The intake mounting flange is unique among V-8 engines because it's positioned inside the valve cover, and thus exposed to oil on top, sides, and bottom. Intake mounting and fit on these engines is sometimes challenging because the head ports, pushrod tubes, gasket flanges, valve cover rails, and distributor all must line up for assembly to be successful.

Exhaust ports are largely outboard of the valve covers on an FE, giving a signature external casting appearance not seen on other Ford engines. Spark plugs are comparatively vertical in orientation in comparison to a 302 or a 460. Factory casting numbers are found facing upward on the flat surface in between the center exhaust ports, and follow the common Ford numbering routine.

Valve spacing and sizing changed depending on the head and application. The common heads used on 390 and 428 engines all share the same valve spacing and similar rocker gear. The high-performance heads had unique wider spacing in order to accommodate larger intake and exhaust valves. In concert with the spacing changes, rocker stands (covered in more detail later) were altered.

At a salvage yard or swap meet, you're likely to find many OEM cylinder heads in similar condition to this example. They often require substantial work and the cost of repairing and improving these can often approach the cost of replacement.

The wide spacing combined with the larger-diameter valves mean that factory 427 heads will physically bolt on but do not function properly on a 390 or 428 block because the valves will hit the cylinder bore when open. This can be remedied to a point by notching the top of each cylinder for valve clearance, but the close proximity of the bore reduces the flow benefit from the bigger valves. Cobra Jet heads can also get very close to the bores on a 390 engine, which is definitely something that needs to be checked.

Factory Cylinder Heads

There's a lot of confusion surrounding the various factory cylinder-head designations on FE engines. With many aftermarket intakes being marketed as "high rise," it only

This is the same head viewed from the intake corner. As you can see, this head has seen extensive service. While there are numerous variations, most non-performance OEM heads can be considered reasonably comparable in terms of their power potential.

made things worse as every parking-lot hot rodder and online merchant is convinced that he has a real high riser on his hands.

The various "riser" designations came from Ford's factory race efforts in the early to mid 1960s. In the beginning, the FE had a single, basic intake port layout, which saw the normal incremental changes as development continued.

High Risers

As factory involvement in racing took hold in the early 1960s, it developed a specialized racing cylinder head that was used first in limited-production lightweight Galaxies and later in the Fairlane Thunderbolt. This race-oriented head had numerous unique characteristics, including a different valve-cover mounting-rail angle and rocker stand heights. But the intake port was the key feature. It was dramatically taller than those on the traditional FE head. This mandated a matching tall intake, which in turn required a lump in the hood–the instantly recognizable teardrop bubble–and the term "high riser" was coined.

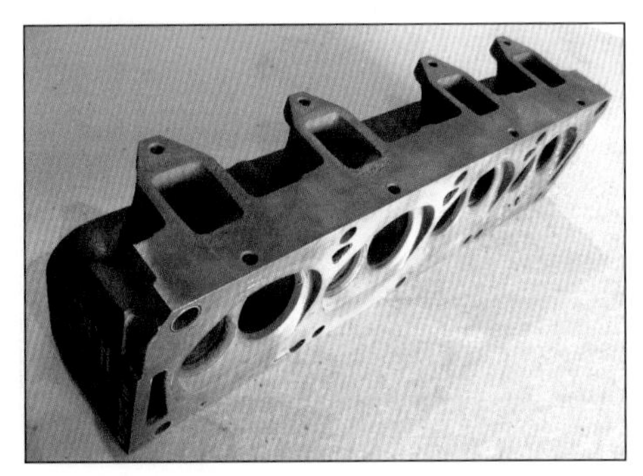

The C4AE high-riser exhaust port heads were fitted to the 427, so these are rare heads indeed. This race-oriented head had numerous features that set it apart from the stock 390 heads, such as different rocker-stand height, and valve-cover angle. The intake port was the key feature and it was far taller than those on the traditional FE head.

The C4AE high-riser head's unique, tall, and narrow intake port and fully-machined combustion chamber are readily apparent.

You can easily see the difference in valve-cover rail angle and rocker pedestal-mounting height between the medium-riser head (left) and the high-riser head (right).

High-riser heads have considerably larger valves than traditional passenger car heads, along with a wider distance between the valve centerlines. Also, a much larger combustion chamber—at around 86 to 88 cc—required a domed piston to achieve high compression. The rocker system for the higher riser at first glance looks like any other FE, but it actually uses a shorter pedestal height in order to accommodate the casting's raised intake port.

High-riser heads were always rare, only came in limited-production specialty vehicles for a couple of years, and have become extremely valuable. You will not find them in a salvage yard or at a garage sale. They were always race parts and most of them led a hard life with extensive modification and repair being common. It is safe to assume that all remaining usable castings are in the hands of either collectors or racers.

High-riser rocker mounting is shown in detail, and once again, notice that the pedestal-mounting height, which is unique to the high-riser.

Low Risers

So, if the new race head was a "high riser," everything else needed a name too, and thus they became "low risers." While this best describes the earlier 390, 406, or 427 performance heads, the low riser designation has become a default term used to describe every non-performance factory casting as well. As a result, there is a dizzying array of head castings that fall into the low riser category. Fortunately for the guy building on a budget, most of these castings will physically bolt on and work on all popular FE engines. The 427-based heads that would cause any valve interference issues are readily identified collector items; a casual builder is unlikely to stumble upon a set.

Most factory low-riser heads are functionally interchangeable, sharing a common port entry, rocker pedestal position, and chambers that are reasonably similar in volume. Some castings are more desirable

The C3AE-D head's intake port and chamber is a classic low-riser design. The port opening is actually larger than the later medium-riser design, but not as tall as the exotic high-riser. The low-riser designation is based on the basic size and floor position of the port. Most non-performance FE heads have come to be considered "low risers" by default.

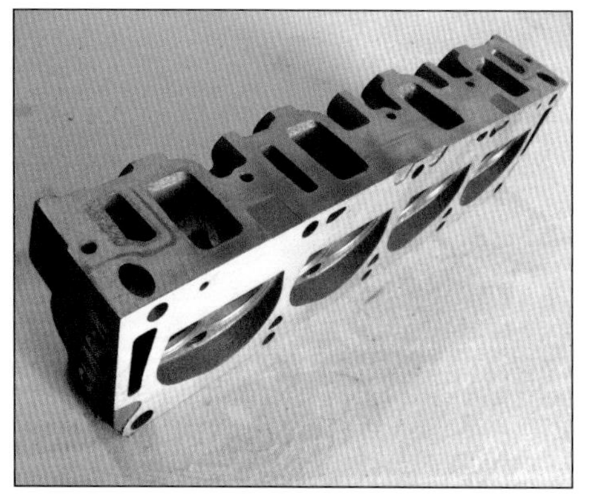

Some builders consider the C6AE-R head to be the best of the standard heads, and this particular one was the last passenger-car head to use a full-size variation of the low-riser intake port. Most subsequent non-performance FE heads from 1966-on had a smaller port opening. Realistically, these are good parts, but nothing particularly special. You can find a number of aftermarket heads that perform better than these, even after modest porting work.

than others, but the differences and advantages are generally modest. Exhaust port position can cause some issues with header compatibility. The volume header manufacturers did not recognize that, although the bolt pattern remained the same, there are two different port heights relative to the deck.

Medium Risers

The medium-riser combination actually followed after the others, and was in part an effort to get a race-oriented 427 into a production vehicle without the hood bubble. The medium-riser heads share the standard head's valve-cover rail and reasonably similar rocker-stand heights. Medium-riser intake ports have a raised floor and a more squared-off port cross section, compared to other FE heads. This was said to result from experience and research showing

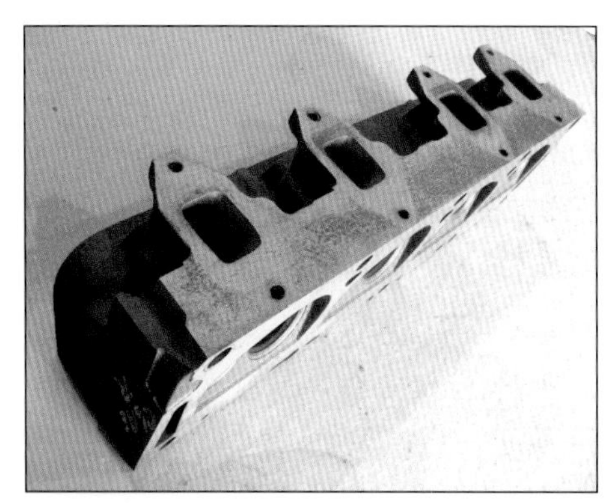

Like other FE heads, the C6AE-R head has an exhaust side with a normal 8-bolt pattern. Heads from 1966-on, including the later truck heads, have the exhaust port in a lowered location while the bolts remained in the same spot. This can make it difficult for some headers to fit because they overlap the port opening causing header leaks and reduced power. Since header manufacturers do not specify which head was used to master their mounting flange, the only way to determine whether any particular headers will fit is to mark and test fit them first.

A C6AE-U GT390 head has an exhaust side with a unique 14-bolt pattern, and therefore it's a special part to fit a few applications. In fact, the heads were originally designed to ease fitment of the exhaust side of the manifold in the Mustang and Fairlane platform. The 14-bolt exhaust pattern only fits the stock manifolds properly and even headers supposedly designed for these cars don't fit the exhaust pattern well.

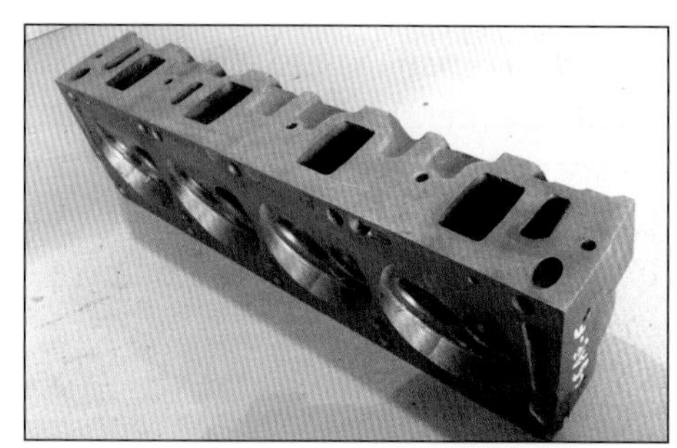

A C5AE-F head shows the medium-riser intake port and the machined wedge combustion chamber. The medium riser was released in 1965 and had a design permitting it to fit under a conventional hood. The valve-cover rail and rocker mounting were also more conventional than the high-riser predecessor. Medium-riser heads have a very efficient port size and shape for the era. As such, they served as the design inspiration for the Edelbrock heads. The port is smaller than the traditional low-riser head and yet flows more air.

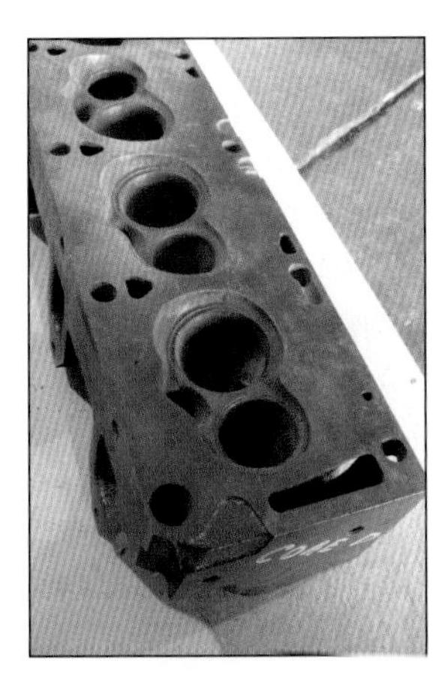

This particular high-performance head, the C0AE-D, was fitted to early high-performance 352s. Note the unique machined heart-shaped combustion chamber. Ford continued to machine chambers during the 427 era, but the heart-shaped concept was largely abandoned for many years. Current race heads have a chamber that, while far more sophisticated, is eerily reminiscent of the old 352 parts.

The C0AE-D head's exhaust-port size, bolt pattern, and port position relative to the deck is conventional for FE engines. It is shared with most 1960s-vintage FE heads, but the head's port size, bolt pattern, and other aspects changed with the Cobra Jet and later truck heads of the 1970s.

The C0AE-D has the large low-riser-size intake port opening and a small-volume chamber measuring roughly 60 cc. It is important to note that the actual chamber volumes of FE heads can vary dramatically from the published data, especially on "as-cast" chamber heads. Machined chambers are normally more consistent, but these being 50 years old they've likely been machined a time or two.

A C3AE-D is a 1963 427 low-riser head, which uses the standard FE exhaust-port location and vertical eight-bolt "up-and-down" pattern.

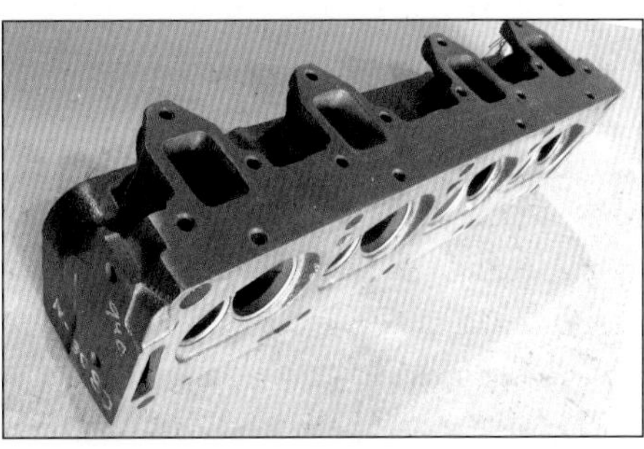

The Cobra Jet heads share the same basic valve cover and rocker parts as the heads on the common 390 passenger car, but these have larger valves at 2.09 and 1.65 inches in diameter. A C8OE-N Cobra Jet head exhaust port has a unique 16-bolt pattern. The upper and lower bolts are in the normal post-1966 FE locations.

that most airflow occurred at the top and sides of a port, rather than along the bottom, and that the reduced port cross section for a given amount of airflow made the head more efficient. While seemingly intuitive in today's context, the move to a physically smaller port cross section was quietly revolutionary considering that every other performance engine, including those from Ford, would still be going down the "bigger is better" path for several years.

The GT 390 heads are often visually mistaken for 428 CJ heads, but they are best used for a numbers-matching project. Designed originally to ease exhaust-side fit in the 1967 Mustang and Fairlane platform, they have a unique exhaust pattern (14 bolts) that properly fits only the stock manifolds. Even headers, ostensibly intended for the application, don't fit the exhaust pattern well. With their normal intake port and standard-size 2.03-inch intake and 1.56-inch exhaust valves, and none of the performance-enhancing features that make a Cobra Jet head desirable.

Cobra Jet heads are a key component in what would be the "last hurrah" for factory-installed FE high-performance engines. It's intuitive to want to categorize these as a "low riser" or a "medium riser," but in reality they are a blend of the two, along with an important feature only found on these—the 16-bolt exhaust flange. A Cobra Jet head has an intake gasket face that matches the low-riser gasket dimensions, but the port itself transitions to the medium riser's general cross section and floor a short distance into the port. The Cobra Jet heads use the same basic valve cover and rocker parts as the traditional passenger-car 390, but have larger valves at 2.09 and 1.65 inches in diameter. The exhaust side of the head was altered to accommodate eight additional horizontally oriented fasteners, easing installation in shock-tower-equipped Mustangs and Torinos. This exhaust face, coupled with the C8OE-N part number, make a Cobra Jet head readily identifiable.

The Cobra Jet was perhaps the most popular high-performance FE engine, and the C8OE-N heads were unique to this package. These heads were a combination of low- and medium-riser head features. As such, the intake port is different than other FE engines while still fitting into the common architecture, having a defined ramp shape to the floor of the port.

Exotic Heads

The tunnel port and single overhead cam (SOHC) are always pulled to the side in discussing FE cylinder heads, due to the dramatic differences in layout and design. The high-riser head should also be included in this list, given the non-interchangeability with common FE components. All three of these are developments from Ford racing programs, and were never in volume-production cars. Only the high riser made it into vehicle production at all; the others were over-the-counter race parts only.

As previously mentioned, the high-riser head has a dramatically taller than the usual intake port, and requires quite a few unique components in order to be installed. The intake manifold is different in both port design and valve-cover mounting rail position and angle. The rocker stands are considerably shorter due to the raised upper surface of the head, again done to accommodate the ports. These characteristics mean that none of the currently available high-volume-production aftermarket intakes or rocker systems will fit. The only aftermarket intakes for these heads come from Dove—a specialty supplier dedicated to low-volume FE racing parts.

Tunnel-port heads are another completely different cylinder head that came from Ford's racing program. Never installed in any production engine or car, these heads were an effort to address the airflow needed in continuous high-RPM-racing applications, such as NASCAR and LeMans. Rather than trying to snake the ports around the pushrods as done in conventional heads, Ford simply put the port where it wanted it and ran the pushrod through it inside of a tube. The ports themselves are very large and rounded in shape. The intake manifolds for tunnel-port heads are unique in port design and bolt pattern. They are very interesting parts with a well-proven record of racing success—but not parts you are going to find in the average swap meet or street car on a Friday night.

The SOHC 427 engine (the so-called Cammer) was perhaps the most exotic package that actually made it into production during the muscle car era from Detroit's "Big 3." It was never sold in cars, but was available over the parts counter—a 1960s crate engine. The huge cylinder heads with single overhead camshafts are the visually overwhelming feature, but the package entailed a lot more than just those. With hemispherical combustion chambers, a solid-roller cam

A C7OE tunnel-port head's exhaust and chamber view is almost normal compared to the radically altered intake side.

The C7OE tunnel port intake side is unusual, with the pushrods in tubes that run through the center of the intake port. This is a rare factory race piece that's difficult to find and very expensive to obtain. The tunnel port has a unique intake-manifold bolt pattern that is not shared by any other FE. While certainly very cool parts to have and run, many modern aluminum heads provide comparable or better performance.

An incredibly rare aluminum factory tunnel port intake is shown, and to validate its rarity, check out the part number: zero.

Aluminum tunnel port head is a factory race piece from the late 1960s, and it's most likely from the GT40 program.

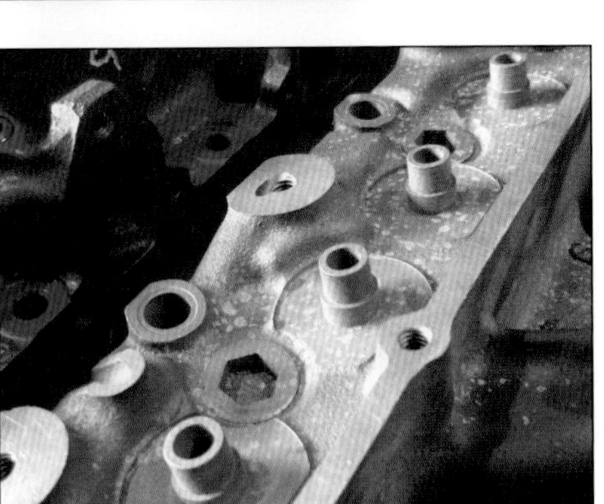

Medium-riser and conventional FE rocker pedestal mounting is shown here. The shaft heights are reasonably similar, although the valve center-to-center distances are different.

From top to bottom, here are exhaust-side heads for the Cobra Jet, medium riser, high riser, and tunnel port. You can easily see differences in the chamber designs and sizes.

From top to bottom: tunnel port, high riser, medium riser, and Cobra Jet chamber side.

From right to left: the intake side of the tunnel port, high riser, medium riser, and Cobra Jet.

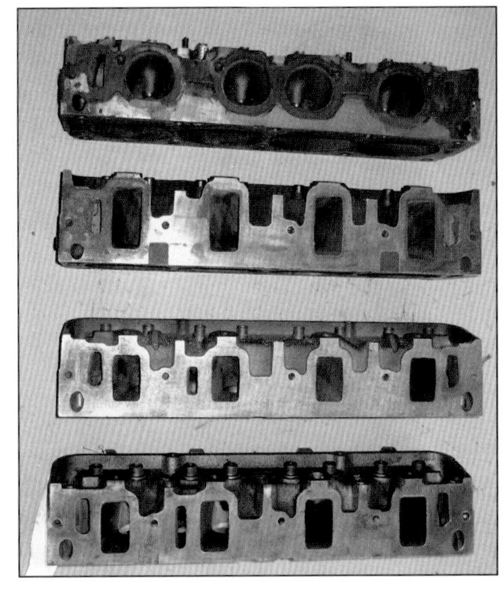

From top to bottom: the side view of the tunnel port, high riser, medium riser, and Cobra Jet intake.

From right to left: the top side of the tunnel port, high riser, medium riser, and Cobra Jet top angle.

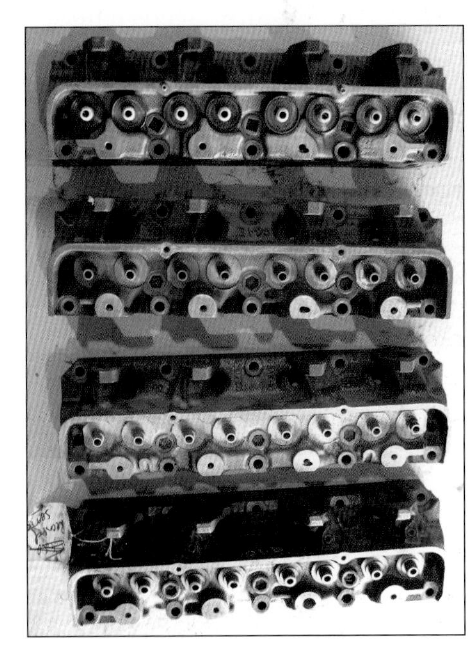

From top to bottom: tunnel port, high riser, medium riser, and Cobra Jet top sides.

Here is a chamber-side comparison of an Edelbrock (top) head and a Blue Thunder (bottom) head. The as-cast Blue Thunder head has a modern heart-shape design that we further modify with a CNC program. The fully profiled chamber enhances combustion, improves flow through the opened valve, and has a reduced timing requirement. More power, better fuel quality tolerance, and greater efficiency are the result.

mounted in each cylinder head, a unique multiple-chain cam drive, plus specialized intakes, front covers, valve covers, and valvetrain, this should almost be considered a completely different engine built around an FE block and crank. The SOHC has recently seen renewed interest, as several specialty suppliers have begun reproducing the components needed to build them from scratch.

Edelbrock Heads

The release of the Edelbrock aluminum cylinder head was key to the mainstream resurgence of the FE engine. While not a watershed design, it is a well-engineered and good-quality part that addressed the market need for a replacement, as the number of economically repairable original parts dwindled. The various Edelbrock heads share a common casting with differences in machining and installed components. The basic head has remained essentially unchanged over its entire production run, with only minor alterations in machining.

The intake port design and gasket face are essentially medium-riser inspired. Most of the factory intake manifolds can be made to work, as can most aftermarket intakes that are designed for 427 medium-riser or 390/428 CJ applications. The exhaust side is offered with either the 428 CJ 16-bolt pattern or the traditional 8-bolt vertical pattern, but the castings are identical. Exhaust port and bolt patterns are at the higher FE factory position.

Combustion-chamber choices are either a nominal "as cast" 72 cc or a machined 76 cc simple wedge. Valve sizing on the complete assemblies are 428 CJ-derived 2.09 inches for the intake and 1.65 inches for the exhaust. The valves and valve job delivered on the assembled version of the Edelbrock head are functional replacements for a 428 CJ with 3/8-inch stems, a 30-degree single-angle intake seat, and a 45-degree exhaust seat.

Valve positions were altered slightly to allow clearance and "bolt-on" use on the base 4.050-inch-bore 390 engines. Rocker mounting is with the standard FE four 3/8-inch-16 fasteners. The factory parts will work—although minor shimming of the rocker arms may be required to accommodate the aforementioned valve-position change. Spark plugs are the smaller 5/8-inch hex, 14-mm 3/4-inch reach design.

In "as delivered" condition, the assembled Edelbrock head represents a good value for the street-oriented enthusiast. Performance is slightly better than a 428 Cobra Jet cylinder head, coupled with better availability and a cost that is reasonably close to the price of repairing and upgrading aged factory castings. Equally important, these heads can be significantly upgraded and extensively modified for high-performance street use and track duty.

As delivered, Edelbrock heads support a street-driven 390 or 428 up to around 450 hp before becoming a constraint. With larger valves, a modest amount of work to the valve bowls, and a professional multi-angle

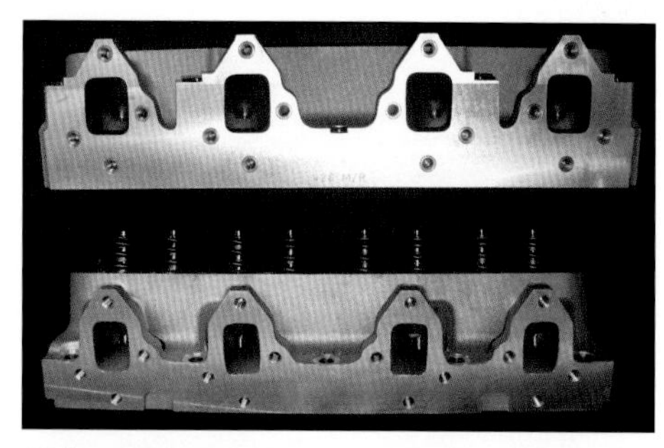

This exhaust-side comparison shows the difference in port height relative to deck surface: Blue Thunder is above; Edelbrock is below. The Blue Thunder head has a .400-inch raised exhaust-port position relative to the deck.

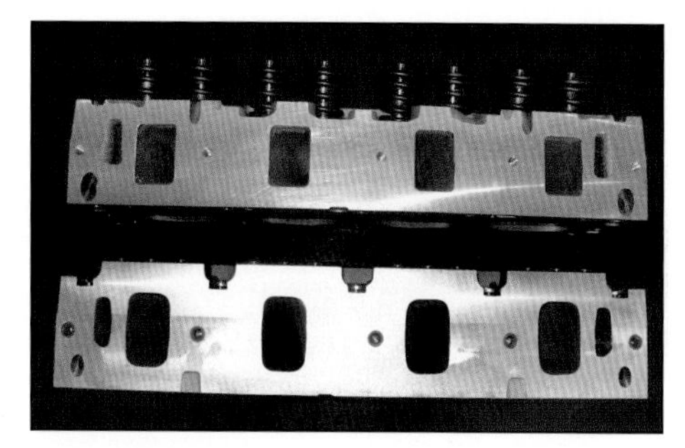

Blue Thunder is on the bottom and Edelbrock is on the top, so the intake flange on the heads can be compared. Both heads place the intake, exhaust, and valve-cover flanges in the stock location. Note the large and round-cornered port openings on the Blue Thunder head, along with the raised rocker-mounting pads.

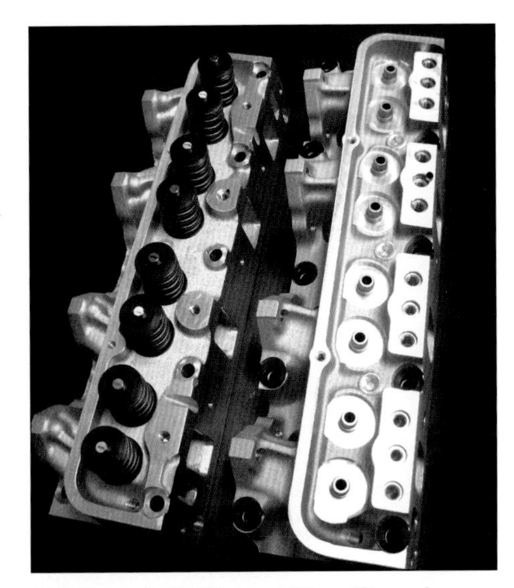

Edelbrock (left) and Blue Thunder (right) heads viewed from above. You can readily see the added rocker-mounting provisions of the Blue Thunder head, while the Edelbrock is more conventional. The Blue Thunder features a high-riser rocker-pad height and allows use of a dedicated T&D rocker system, so a more aggressive cam package can be used without compromising strength. An additional series of eight 7/16-inch bolts mount the T&D rocker shaft system, which gives the system great strength.

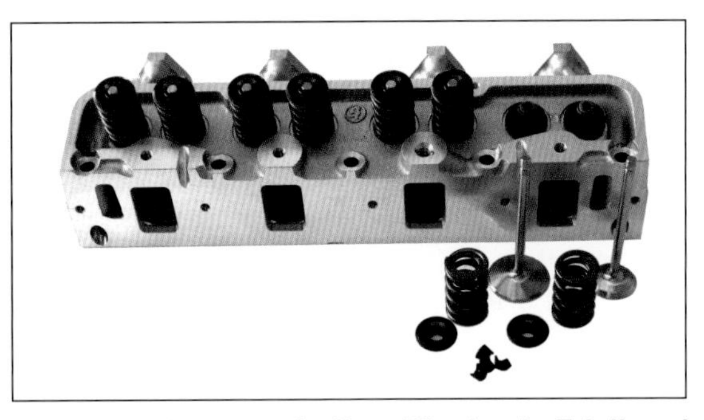

Modified Edelbrock FE head assembly with bigger valves and better springs. The intake valves are 2.200 inches in diameter with a smaller 11/32-inch stem. This package, when combined with a CNC-generated race-quality valveseat profile, delivers a 7-percent increase in flow. The basic Edelbrock springs are a single design with a damper. They are swapped out for a set of double springs that allow larger cams and greater longevity.

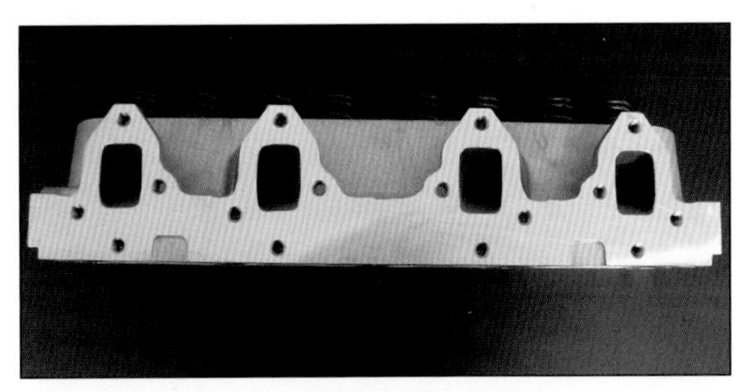

Edelbrock FE heads have a variety of exhaust-flange bolt patterns. The 72-cc version of the head has the Cobra Jet–style 16-bolt exhaust pattern, while the 76-cc head only has the traditional 8-bolt vertical FE pattern. The 16-bolt Cobra Jet–style head provides more performance potential, because it allows you to CNC tailor the chamber shape—and it's easier to install headers on a Mustang or Fairlane.

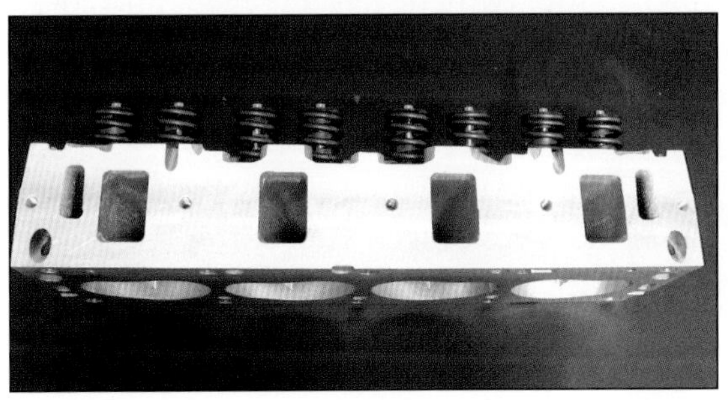

The Edelbrock head intake flange is compatible with a common medium-riser-size runner and with the 1247 Fel-Pro gasket shape.

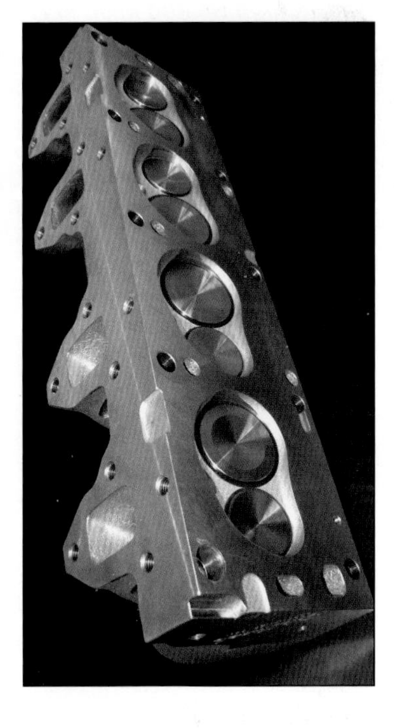

Edelbrock heads feature a lower corner and chamber. Edelbrock offers a machined 76-cc chamber as well as this cast 72-cc version.

Here is the combustion chamber on a modified Edelbrock cylinder head. The detail shows the multi-angle valve job and the blended bowl area above the valve. These combustion chambers are easily shaped or modified for high-performance street use and track duty.

valve job they will suffice to beyond 500 hp. The 390-based engines do not accommodate a larger exhaust valve, but a 2.200-inch-diameter intake does clear. Extensive port work delivers a head capable of supporting a 482-ci engine with more than 600 hp.

The Edelbrock head oil feed and rocker mount is similar to the factory medium-riser layout. The rocker-mount position is correct for the factory adjustable 1.76:1-ratio rockers. The factory assembled heads allow valve lift up to a maximum of .600 inch. In comparison to the 390 and 428 heads, the intake valve has been repositioned away from the bore centerline.

This view of the Edelbrock head's upper corner shows the conventional FE architecture for bolt locations and rocker mounting. The Edelbrock head is designed as something of an upgraded service replacement for 390 and 428 engines, and can support more than 450 hp as cast. These are popular and reliable heads for many high-performance street engines.

Blue Thunder Heads

The Edelbrock head can be considered a bolt-on, cost-effective alternative to repairing factory castings and a part that has good performance potential. By contrast, the Blue Thunder heads approach the FE market from the other direction, bringing race-inspired features, and significant design revisions to a streetable package. The Blue Thunder heads focus on the more serious engine combinations, with larger displacements of 450 inches or more and RPM expectations beyond 6,500. These heads are delivered with valve seats installed but not machined. Guides are included but not installed. The consumer is responsible for correct machining, component selection, and assembly.

Blue Thunder offers multiple heads from a common external casting, which has evolved and been upgraded over the past few years. Current castings have a 16-bolt Cobra-Jet exhaust pattern, which is raised .400 inch relative to the head's deck surface. The current offerings include heads with either a high-riser or a medium-riser intake port, neither of which truly resembles the factory parts that share those names.

The current Blue Thunder heads require nine "long" head bolts (or studs) and a single "short" one in the lower center position. Factory castings, in comparison, use five of each length. ARP offers a stud kit specifically for the Blue Thunder heads. Each head-bolt position has a pressed-in hardened washer to prevent galling or distortion. All fastener holes have stainless-steel thread inserts.

Rocker mount holes are provided for multiple systems, both factory and aftermarket. Four 3/8 -16 holes are in the normal FE locations, along with eight 7/16 -14 holes that are aligned with the valve centers, as is common in most other engines. Rocker-mounting positions are raised and require the shorter high-riser pedestals for factory-style systems. The thread inserts are recessed to allow milling of the rocker-mounting surface if needed for geometry adjustment. TD Machine offers a bolt-on rocker system that utilizes the additional fasteners and "through-the-pushrod" oiling to address weaknesses inherent in the factory setup when run in race applications with high spring pressures.

The more popular medium-riser casting uses a large CJ/low-riser-sized port opening with the port floor set at the Ford medium-riser position—descriptively closer to being a raised-port Cobra Jet head. The as-cast port is large and well shaped, and can be run unaltered in 482-ci engines making more than 700 hp. The large port opening does limit intake manifold selection—only the Blue Thunder

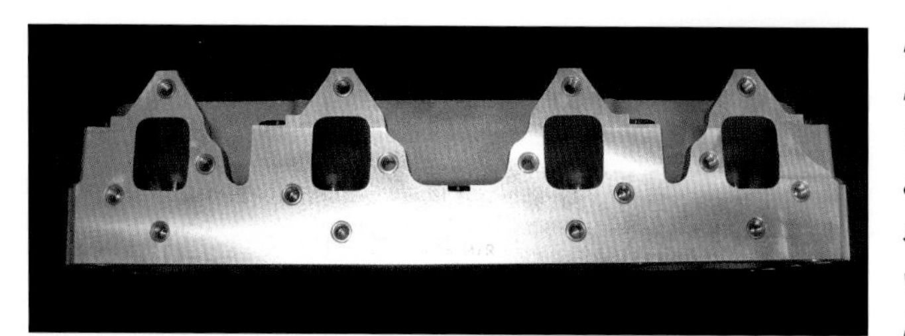

Medium-riser and high-riser Blue Thunder heads share a common exhaust side. The older version of the heads had the 8-bolt FE pattern, while newer castings now require longer head bolts and have the 16-bolt Cobra Jet pattern for improved clamping force and high-performance use.

The Blue Thunder high-riser head is designed to be CNC ported, and it is shown here on the intake side. The Blue Thunder high-riser heads come with a fairly small port.

Here is a comparison between the Blue Thunder and Edelbrock medium-riser cast port openings using the gaskets as a reference. You can readily see the difference in port opening size and shape. The Blue Thunder is noticeably larger and rounded off compared to the sharp-cornered Edelbrock.

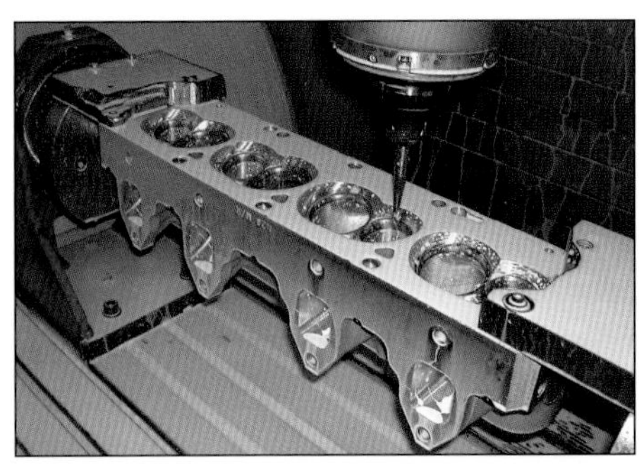

The combustion chamber on this Blue Thunder head has been profiled. The desired port and chamber shape is digitized, and then tool paths are programmed allowing the CNC mill to effectively duplicate the model.

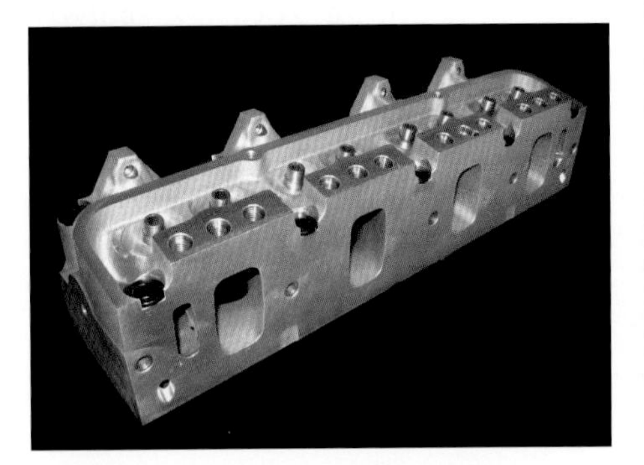

The Blue Thunder medium-riser heads used to come with a small port, but are now shipped with a much larger port opening that is effectively ready to use. They still require a valve job and components to perform at their best.

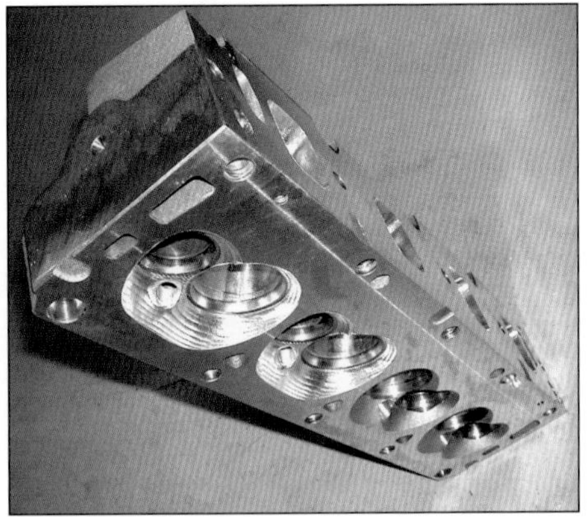

Machining a CNC chamber. This is a first rough cut, which will be followed by a fine cut. CNC machining allows quick and exact duplication of a developed and proven port design or chamber profile.

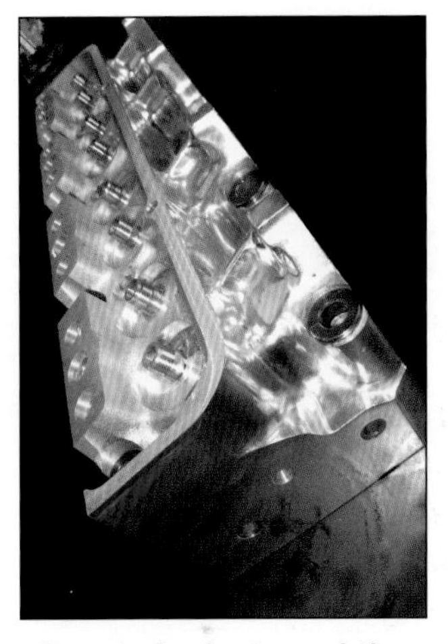

A completely CNC-ported and polished Blue Thunder high-riser head is a pretty impressive piece to look at—and they run as well as they look. This head was installed on a 520-inch, 12:1 compression, dual-quad-equipped engine that made more than 770 hp.

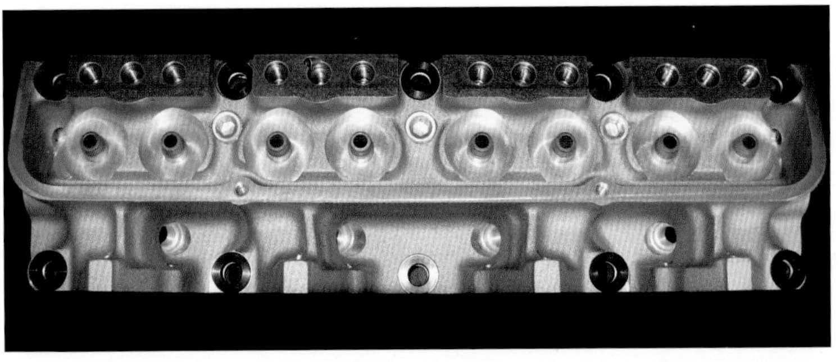

These Blue Thunder pieces are race-oriented heads, and the differences between these heads and the traditional FE parts are readily apparent. Since these are heads intended for engine builders and specific high-performance applications, valve seats are installed but not machined. Guides are included but not installed, so the builder/owner needs to correctly machine the guides, select the correct components, and assemble the heads. These heads easily accommodate 450-ci-or larger displacement engines that can rev more than 6,500 rpm.

dual-plane intakes and the Edelbrock Victor have enough material to match the runner size without welding.

The Blue Thunder high-riser port is intentionally cast small, so that the engine builder can extensively port it by hand or CNC machine. As delivered, the head has a port opening roughly the same size as a Ford medium-riser head, but with the roof set to the high-riser position—descriptively more like a raised-port medium-riser head. In professional port work I have seen, examples of this casting reach 400 cfm in airflow—a range previously impossible with traditional FE castings. Blue Thunder chose to adopt the unique factory high-riser valve cover rail angle, which means intake selection is very limited, but the "Paquet" port intakes from Dove work well. At this build level, a fabricated sheetmetal intake is also a viable option.

Dove Heads

Although beyond the scope of this book, I know several professional racers who run specialty heads cast by Dove. These tend to be very-low-volume semi-finished castings, and require considerable preparation and effort to get them into a "race-ready" condition. But Dove does offer a wide array of port and chamber casting options, and the parts do warrant consideration for serious race-engine builders.

Here is a Dove aluminum high-riser head with an altered exhaust face. As a small supplier to the race community, Dove offers a variety of highly specialized parts. Dove manufactures numerous parts on a limited-production basis, and has cataloged exotica such as an aluminum tunnel port head (SK-37080) and a cast-iron tunnel port (C70E-6090-K) for large-displacement engines and specific racing applications.

The intake side of the Dove high-riser head is shown. Dove offers both a factory high-riser port, which is tall and narrow, as well as a raised and wider Paquet port.

A Dove aluminum high-riser head with common exhaust port design provides an interesting comparison. The Ray Paquet–inspired raised-port high-riser head design provides exceptionally-high flow characteristics from a very broad power band.

Flow Data

It has become common to use cylinder head port flow as a primary key in determining a head's power potential. While certainly a contributing factor, flow numbers alone do not really give a complete picture. Combustion-chamber shape, plug location, valve position relative to the bore, air speed, runner cross section, and mixture motion are among the innumerable variables that a simple flow value does not address. Nonetheless, without some relevant flow numbers, this chapter would not be complete.

The following is a good selection of flow data for the various FE cylinder heads. This information should be considered directional rather than "absolute." It seems flow numbers vary from location to location depending on bench brand and model, as well as operator variables. All of this information was derived from testing using a SuperFlow 1020 bench at the industry standard 28 inches of depression. The exhaust flow data is without an extension pipe, a common industry practice that would raise the numbers by perhaps 10 percent.

The values shown are meant to be illustrative, and your chosen shop's equipment may yield better or worse numbers from similarly modified heads.

First up are the Edelbrock FE heads. The "stock" numbers are for the 72 cc heads with their 2.090-inch and 1.650-inch valve combination—as delivered with no changes. The "mild rework" heads have 11/32-inch stem and 2.200-inch intake valves, 1.710-inch exhaust valves, some bowl blend work, and a really good multi-angle valve job. The CNC-ported heads use the same valves and valve job. But these have completely recontoured intake and exhaust ports with considerable material removal, along with a profiled combustion chamber. They are intended to be as aggressive as possible while retaining compatibility with production intakes and gaskets.

And the Blue Thunder heads, with the Edelbrock CNC ported data, are included for comparison. Not that the two Blue Thunder heads use essentially the same exhaust port, with the

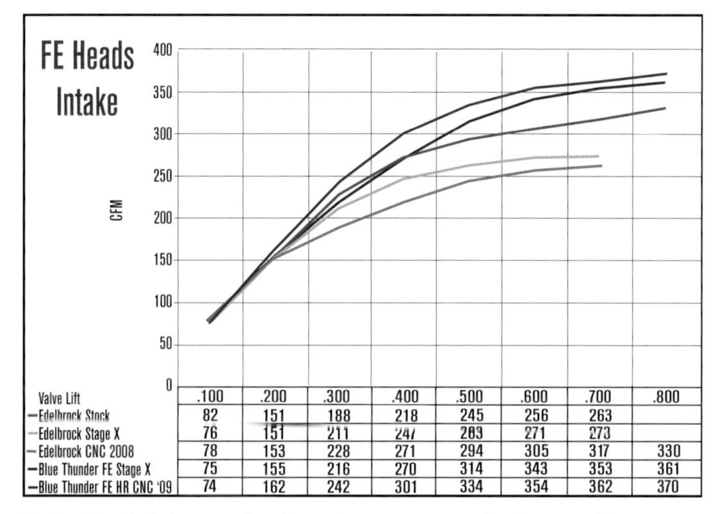

Valve Lift	.100	.200	.300	.400	.500	.600	.700	.800
Edelbrock Stock	82	151	188	218	245	256	263	
Edelbrock Stage X	76	151	211	247	283	271	273	
Edelbrock CNC 2008	78	153	228	271	294	305	317	330
Blue Thunder FE Stage X	75	155	216	270	314	343	353	361
Blue Thunder FE HR CNC '09	74	162	242	301	334	354	362	370

Data for intake and exhaust was generated on a Super-Flow 1020 flow bench at an industry standard 28 inches of depression.

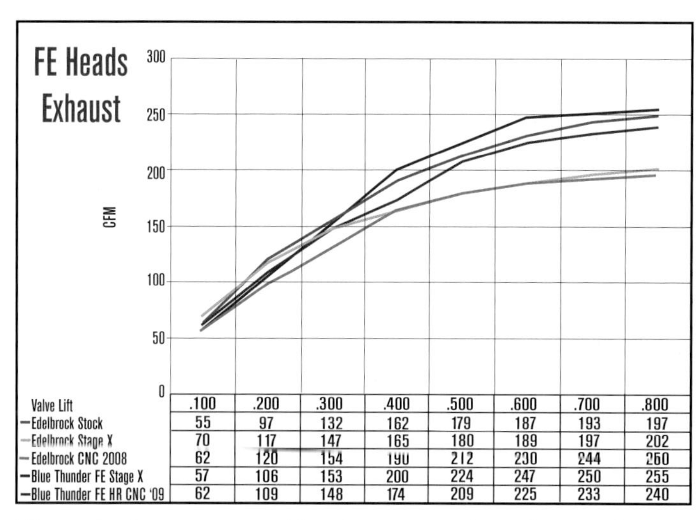

Valve Lift	.100	.200	.300	.400	.500	.600	.700	.800
Edelbrock Stock	55	97	132	162	179	187	193	197
Edelbrock Stage X	70	117	147	165	180	189	197	202
Edelbrock CNC 2008	62	120	154	190	212	230	244	260
Blue Thunder FE Stage X	57	106	153	200	224	247	250	255
Blue Thunder FE HR CNC '09	62	109	148	174	209	225	233	240

These tests were run without an extension pipe on the port, which adds about 30 cfm to the values shown.

same operator using the same flow bench, but on different days, thus the variable flow data serves to point out the pitfalls inherent in flow testing.

The medium risers had a CNC-profiled combustion chamber with 2.200 and 1.710-inch exhaust valves. They had a very good valve job and bowl blending, but the ports themselves are untouched.

The CNC-ported high-riser heads tested are a maximum-effort race set, with 2.250-inch intake valves and 1.750-inch exhausts in a profiled chamber. They deliver the shown flow increase, yet still have a smaller port cross section than do the "as-cast" medium-riser heads. They've been proven to deliver more horsepower and torque despite the aforementioned smaller ports–proof that flow numbers are not the only factor influencing head performance and selection.

What to Choose?

When choosing the correct cylinder head for an FE package, you must weigh the options between availability, cost, performance potential, and ease of installation in the engine and the vehicle. Each head has advantages in certain applications, and there is a fair amount of overlap where multiple choices will be successful.

For the lowest-cost build with modest power goals a common factory production head suffices, and performs even better with good valve work and attention to detail. The castings are nearly free in many cases. Cost of the machine work varies dramatically, depending on condition, but almost always is lower than purchasing new replacement parts. If you have connections with a machine shop, the price may be even lower.

If you are building a restoration piece or a class-legal race car, the casting choice is already made for you. The rules and your checkbook dictate repair and alterations/porting. Extensive porting on rare factory castings is challenging and expensive.

If you are building a mild street-performance car or truck, the Edelbrock heads are good options from "a dollars-per-horsepower" perspective. Out of the box, they'll support around 450 hp. Modest upgrades have the average combination handled even beyond the 500-hpmark.

For a higher-powered street- or strip-type build, the choice can get tougher. Extensively ported Edelbrocks fit well within common engine bays and accept readily available valvetrain systems. The Blue Thunder heads usually make more power. But they can be more expensive and more challenging to fit into the overall package. Headers can be tougher to install due to the raised exhaust—causing potential shock-tower interference on Mustangs or side-pipe position and layout problems on a Cobra. The best valvetrain to use on them is the expensive TD system. But the added expense and effort has real benefits in terms of power potential and durability. As the desired power level increases, along with the installer's ability and willingness to address the installation challenges, the Blue Thunder head emerges with some significant advantages.

CHAPTER 7

Rocker Systems

The factory non-adjustable rocker system is a simple system that works for common street use. The end rockers are unsupported, which is fine for a stock rebuild or mildly built street engine. However, if a larger, aggressive cam and stiffer valvesprings are installed, the rocker system is susceptible to breaking.

Ford FE engines (with the notable exception of the SOHC) use a single shaft-mounted rocker assembly per cylinder head. This shaft is attached to the head with four pedestals and 3/8-16 fasteners. Factory-style rockers, whether original equipment or aftermarket replacements, are steel or ductile iron with either a ball or cup pushrod end and a radiused contact for the valve.

Factory Rockers

Factory rockers come in adjustable and non-adjustable designs. The latter is far more common and is used in most OEM hydraulic lifter FE engines. Adjustable-style factory FE rockers have an interference thread-adjuster screw on the pushrod end and use a cup-type pushrod, as compared to the ball end used with the non-adjustable rockers. The interference thread is designed to remain in position when adjusted but often loosens in service after many adjustments have been made.

The ratio for adjustable rockers is normally quoted as being 1.76:1, while the non-adjustables are referenced as having a 1.73:1 ratio. The rockers ride directly on the shafts with no bushings. If you see a set with bronze bushings, they have been reconditioned. While bronze is likely a better wear surface than steel alloy, the cross section and strength of the rocker itself has been reduced and it is more prone to break.

Factory rocker assemblies use springs between each rocker to keep them located on the shaft. The end rockers each have a spring washer, flat washer, and cotter pin for retention.

The shaft-mounting pedestals on most FE engines are cast aluminum with a round passage for the shaft to slide through. On certain 427 performance applications, the pedestals are steel with a horizontal split in the shaft passage, allowing the pedestal to clamp onto the shaft. The additional clamping and support is important. The factory shaft setup has a tendency to fracture at the head-mounting hole because the end exhaust rocker is unsupported. The mounting hole on the factory pedestals is often larger or oblong in the section below the shaft to allow more oil to reach the shaft. The pedestals used in high-riser and tunnel-port applications are considerably shorter than the more common parts and cannot be interchanged.

Shafts are gun-drilled steel and have holes for the mounting fasteners as well as for oiling. Oil runs up through the second pedestal on each head, through the center of the shaft, and out the oil holes to the rocker. If there are oil holes on only one side of the shaft, they must face down, toward the cylinders during installation. Factory shafts have holes for cotter pins and cup plugs at each end. The cotter pins serve double duty—retaining the end exhaust rockers, and adding insurance that the cup plugs stay in position.

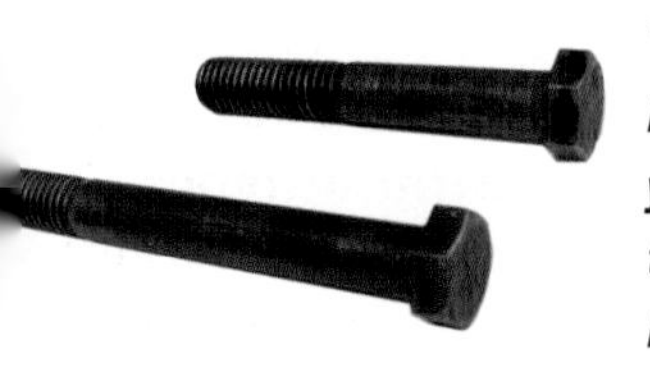

You need to be certain you're using the proper length bolt to prevent damage when installing the OEM rocker bolts. A bolt that's too long does not properly seat in the head. Instead, it bottoms out and gives a false torque reading. This can lead to catastrophic valvetrain failure.

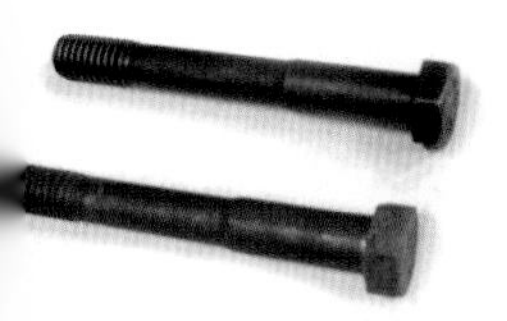

There were different-length OEM rocker bolts depending on the head and rocker pedestal used. The reduced-shank-diameter bolts such as these are used in the oil-feed location. Note the length differences.

Upgrades to the Factory System

The original FE rocker system is considered a weak point, and should be upgraded to match the specific high-performance build. Installing adjustable rockers is the most common initial upgrade so the engine is compatible with an aftermarket cam, modified or milled block, or heads.

A billet rocker stand set with end supports offers a good amount of insurance against shaft breakage at a modest cost. If you're building an FE with an aftermarket camshaft having over .500-inch lift or more than 350-psi valvesprings, I strongly recommend using end supports to prevent shaft breakage.

Detail of the popular Dove roller rockers, shims, and spacer set is used for valve tip alignment. Made of heat-treated aluminum, these parts can handle valve lift up to around .700 inch, and are frequently used for a variety of applications, including street performance and bracket racing.

Replacement or reproduction adjustable rockers are still available from multiple suppliers, and are a simple slide-on replacement for the factory non-adjustable parts. The other change is required for the conversion is the use of different pushrods. Non-adjustable rockers use a common pushrod with a "ball" shape at each end; the adjustable ones use pushrods with an upper cup and lower ball at the lifter end.

Improved versions of the factory adjustable units were available from Crane (now out of business) and Iskendarian. These have flats machined into the rocker body, a longer adjusting screw, and a locknut—as opposed to the problematic interference threads found in factory parts. Aluminum roller tip rockers are a viable option as well, with costs being fairly similar and the advantage of reduced valve guide wear. Horsepower differences are nominal. We recommend upgrading to the aluminum rollers at the 500-hp or 6,500-rpm levels.

Replacing the factory rocker spacer springs with aluminum bushings is an easy and inexpensive upgrade. If you over-rev the engine, the valves can "float" and the pushrods may bounce right out of the cup on the rockers, resulting in considerable damage. Keeping the pieces together with the spacers

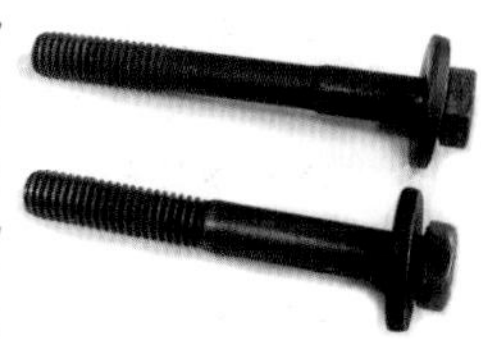

These OEM rocker bolts are what you should expect to find in the head. There should be a single longer, reduced-shank-diameter bolt for the rocker-oiling location and three shorter, standard-shank ones per head.

reduces the amount of carnage in the event of a problem. It's cheap insurance.

The next upgrade should be in material and end supports for the rocker shafts. The factory setup has four 3/8-inch-clearance holes in each shaft for the mounting fasteners to go through. The end exhaust rocker arm is unsupported, and therefore it's fairly common for the shaft to fail at the mounting hole under severe use. Heavier-duty shafts, made from better material with a reduced inside diameter, offer some measure of increased durability.

Multiple suppliers offer systems to provide enough end support. These range from a basic set of stands with end support added, to a sophisticated system using the outer/upper head bolts as an added rocker mounting point. The most common and least expensive is a U-shaped pedestal at each end of the assembly, which serves to capture the shaft and reduce flex. These work quite well in milder applications, and are a good option for budget-oriented builds. I've frequently run end-stand-equipped standard valvetrain systems at 6,500 rpm as long as open spring pressures are below 600 psi.

Aftermarket Valvetrain Systems

Buying each rocker system part individually costs about as much as buying a complete rocker system. Therefore, buying a complete system is often simpler and more cost effective.

Because all the parts are designed to work together, the systems offer better performance and compatibility than separate parts assembled as a complete system. On the good ones, the geometry is right, the spacers and shims (if required) are in the correct positions, and the package has been proven to install and function with a minimum of added effort.

Popular systems follow one of two strategies. They are either an upgraded package that shares the factory-design architecture and features a single shaft with spacers and stands or they are a complete design departure from the factory stuff, with multiple shafts and an adapter mounting plate.

For example, the kits from Dove, Harland Sharp, and Erson follow the first approach. They use the four factory 3/8-inch fasteners to attach the rocker assembly to the heads and require no modification to the cylinder heads for installation. The Erson rocker system is my personal favorite. Erson combined two of the pedestals, an end support, and a spacer into a single aluminum block, simplifying the system considerably while adding rigidity and strength.

In comparison, the system by TD Machine Products uses a mounting plate and four individual shafts, with two rockers per shaft. Five long upper-cylinder-head bolts (included in the kit) fasten the mounting plate to the head for a much more robust assembly than stock. However, if you use the TD setup on a factory or Edelbrock cylinder head, you must machine down the factory-style rocker mountings level with the upper row of head bolts. Note: This renders the heads unusable with any other rocker system—a significant commitment. The T&D rocker system used on Blue Thunder heads is a different story because it uses the additional eight mounting provisions included in the cylinder head—it's quite possibly the most durable FE package available.

Dove FE heavy-duty rocker shafts are more durable than stock parts, but should still be used with an end-support system. On an FE the end exhaust rocker is unsupported and prone to flex. An end stand provides a measure of added support and prevents failure.

Dove offers aluminum roller rockers for 352, 390, and 428 FE engines, and can build special application rockers. These FE rockers are slide-on replacements for adjustable or non-adjustable factory parts.

Installation, Alignment and Geometry

The FE is unique because of the pushrod arrangement. Pushrods install through tubes cast or machined into the intake manifold. It is far easier to measure for pushrod length and check for geometry issues before installing the intake because the rocker system still needs to be removed before the intake is mounted.

The key items covered here are: rocker-shaft height relative to the heads, adjuster settings, and rocker-arm position on the valve tip—both lateral and in contact pattern.

There are multiple schools of thought on the optimum rocker

geometry. One philosophy focuses on minimizing the width of the contact pattern on the valve tip, another concentrates on getting the rocker parallel to the valve at the mid-lift point, and yet another targets maximum possible lift. The TD rocker system provides a gauge to measure the preferred pivot position, but you'll need to test fit and measure the other systems before making a trial run on geometry. The process of measurement and inspection is equally, if not more, important than the chosen methodology or theory because all methods yield similar results on street or moderate strip use engines.

For this part of the discussion, let's assume that valve length and spring installed heights have already been pre-determined because it's a key aspect of the cam selection process. On an FE, the pivot point of the rocker is somewhat "fixed" in comparison to those engines using ball pivot rockers. The only way to change the pivot geometry is to alter shaft height relative to the head or to move the mounting holes, either of which is a major job. The factory-installed spring height is 1.82 inches. If you are increasing this to accommodate larger cam lifts, you need to make a commensurate increase in rocker-shaft stand height with shims or custom stands, so the correct position in the rocker's arc of travel is maintained.

Minimizing the width of the contact pattern probably produces the best reward on a factory-style rocker system where scrubbing occurs as the rocker travels across

Factory FE rockers are located on the shafts by springs and are known to pop out of position at high RPM. Aluminum rocker spacers replace the factory springs and help to keep the rockers centered on the valve tips.

The Erson rocker system uses the unified pedestal and spacers. You must use the supplied small-diameter AN washers when installing the rocker kit so adequate clearance is attained. In addition, you may need to notch the bottom of each pedestal stand to ensure there's enough clearance if you are using ARP head studs.

Offset T&D rockers allow room for wider intake ports in race applications. Offsets of 1/4 inch or more are possible on the intake side.

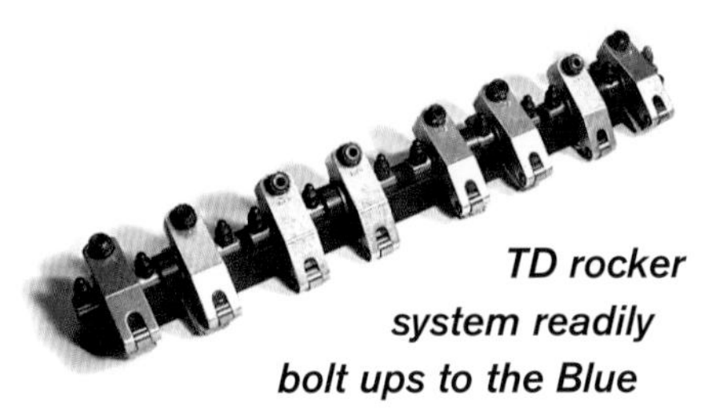

TD rocker system readily bolt ups to the Blue Thunder heads. The rocker system uses the head's additional 8 rocker-mounting holes for greater rigidity and reduced chances of fastener failure.

This is the TD rocker system for Edelbrock heads. Longer upper head bolts are required (and provided) to install the system on the Edelbrock heads, and you need to recognize that head studs are not compatible with this system.

Picture showing a rocker tip properly centered on the valve and some that are off to the side. For proper alignment, rocker tips should be centered over the valve. At the very least, ensure that the roller tip is completely on the valve tip and not off to the side.

Checking and adjusting rocker alignment with Erson rockers is similar to that of other FE valvetrain combinations, but the shims are unique due to the smaller-diameter shafts.

This roller rocker is properly aligned and shows proper valve-tip contact.

To achieve correct valve-tip alignment, the rocker on the right side should be shimmed to the left a bit. Depending on your rocker setup, you may need a variety of shim thicknesses in order to align it.

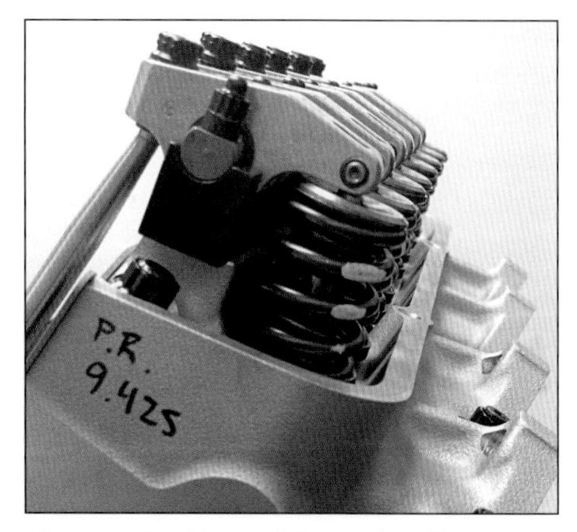

You are looking at the valve-tip position on TD rocker Blue Thunder heads. Note the straight line through the roller axle. Proper geometry reduces the load on parts and ensures durability at high RPM, and under high spring loads.

the valve tip. Reducing the travel intuitively reduces side loads and improves guide life. But on an FE the valve contact pattern cannot be easily altered, assuming that the shaft pivot height has been corrected as necessary. If the contact pattern is off-center from the valve, try a different set of rockers. A rocker that is too short or too long, as measured from the shaft centerline to the contact tip or roller tip, cannot be compensated for. This problem was evident with the Crane rocker system–where the rocker arms were not dimensionally correct. Unfortunately Crane went out of business before the issue could be addressed.

Given the difficulty of altering the contact pattern on an FE, I consider it something to check. But I put a greater focus on centering the arc of travel and avoiding component interference.

In comparison, side-to-side alignment of the rocker tip relative to the valve can be easily altered on an FE engine. A quick visual inspection will show you whether a problem exists. The rocker tip should be centered on the valve tip when viewed from the side. In most cases, you can make an alignment correction by simply adding and relocating the shims that are alongside the rocker in question.

It is very common to find valvetrain-component interference on these engines. The pushrod upper cups hit the underside of the rockers if the pushrods are too long. In addition, factory rocker arms hit oversize valvespring retainers on the underside of the rocker. Factory rockers can also have the shaft surround come into contact with the pushrods on high-lift cams. And it is very common to have the pushrods hit on their "tubes," which are cast or drilled through the intake manifold. All these areas need to be carefully checked and any contact eliminated before final assembly.

Installation

When installing the traditional FE rocker assembly, be very careful not to excessively bend or distort the shafts. The single-unit structure of the rocker assembly means that, no matter what, you will be compressing some of the valvesprings while putting things together.

Carefully and slowly install components to reduce the possibility of damaging the shafts or the mounting fasteners. Get all four bolts started, then move from one to the other and tighten them about a half turn each time. The idea is to keep the shafts as straight as possible and to minimize excessive loads against any single fastener. Be absolutely certain that the bolts have enough threads to tighten without binding against the head or being too long. Fasteners that are too long bottom out and reach the torque value without clamping the pedestals and breakage results. I highly recommend using studs instead of bolts. It makes installation far easier and virtually eliminates any chance of broken or stripped-out mounting holes in the heads.

Note: If you are using the Erson setup, you must use the small-diameter AN washers supplied with the rocker kit. Normal washers, including the original factory parts interfere with the rocker arms. Second, if you are using cylinder-head studs, add a clearance notch at the bottom of each Erson end-stand pedestal. Otherwise, the extra threaded portion of the end head studs hits the pedestal.

When mounting a TD setup to common FE heads—original iron or Edelbrock—you first must have made the machining alterations to the head castings as described in the rocker's instructions. The rocker mounting cradles are then assembled to the heads with the longer head bolts that are included in the rocker kit. Head studs cannot be used with this system. Mounting a TD system to Blue Thunder heads is a straightforward, bolt-on affair using the provided fasteners. It's important to verify the correct orientation of the shaft recess in its cradle; it is slightly offset with the narrower "ledge" facing the intake side of the head.

Measuring for Pushrod Length

The first part of measurement for pushrod length on an FE requires that you set up the lash adjuster. Try to keep only a few threads exposed below the body of the rocker arm. You should never put a side load against a sharp-edged item, due to the risk of breakage. A rocker with the adjuster turned way down positions the cup far away from its oil supply, and also puts the pushrod through a greater arc of travel relative to the intake manifold, thus increasing the likelihood of interference.

You do not want to get the pushrod too close to the rocker though, especially with the FE-specific cup-style pushrods. A pushrod cup hitting on the rocker body at peak lift will cause eventual valvetrain failure.

All of these things may be sacrificed when necessary to correct for other more critical interference issues, but at least then it is done with knowledge of the risks involved.

With the rockers torqued in place, the adjuster in proper position, and the cam on the lowest point in its profile (the base circle), use an adjustable checking pushrod to determine length. The checking pushrod can either be purchased or made from an old pushrod and some threaded rod. On an FE using cup-style pushrods, it is important to verify the proper cup diameter is used during measurement. FE engines use a 3/8-inch-diameter cup while many other engines use a 5/16-inch-diameter cup. Even without a cup-style checker, you can still make an accurate measurement using a ball-end checking pushrod by measuring from the bottom of the adjuster ball. When doing this, you need to specify that you are using "bottom of cup" dimensions; the pushrod company will know exactly what you are saying. With all the clearance between the lifter and the rocker adjuster lightly taken up, remove the checking pushrod and measure it with a 12-inch dial caliper.

If you are running a hydraulic cam, add the desired lifter preload to the measured length. If you are running a solid cam, theoretically, you subtract the desired lash. While preload is measured directly at the lifter, lash is measured at the valve; a .020-inch lash is only .011-inch on the pushrod side due to the 1.76:1 rocker ratio. Because valve lash is even less when cold, the dimensional difference is often very small. A common value is .009 inch cold and thus only .005 inch at the pushrod—approaching the point of being insignificant because the adjusters go .040 inch in a single turn.

Checking the Intake Clearance

With the rockers, stands, and proper-length pushrods chosen it is time to test fit the intake. (Intake installation is covered in Chapter 10.)

The FE is unique among V-8 engines with the pushrods running through cast or drilled passages in the intake manifold. With mild valvetrain combinations, there are normally no problems. But as the cam gets larger and the package gets more aggressive, it is really common to have contact between the pushrod and the manifold. Usually this is a "rub" on the carburetor side of the passage due to the increased arc of rocker travel. The use of offset rockers to permit larger, straighter intake ports also creates side-to-side interference potential.

Test fitting the intake and checking with a flashlight and machinist's dye are the only ways to verify that all interference has been eliminated. It is common to grind on a few passages, and not unusual to find numerous ones needing attention. In addition, it is a fairly common procedure to redrill the openings to accommodate larger-diameter thin-wall-tubing inserts for race engines. On an FE, I strongly recommend test fitting the intake and pushrods before "glueing" components together with any kind of sealant. Therefore, multiple test intake installations are the norm rather than the exception on these engines.

CHAPTER 8

CAMSHAFTS AND LIFTERS

FE flat-tappet cams are ground on iron cores. Most have the necessary oil grooves on journals number-2 and -4 to provide rocker lube to the valvetrain of side-oiler blocks. However, grooving the cam tunnel on the block makes this unnecessary.

Camshaft selection for the FE engine is similar to other cam-in-block pushrod engines. You need to decide exactly what you intend to do with the engine, what you are expecting out of it, what vehicle you are installing it in, and then have a "heart-to-heart" talk with your checkbook.

Among the crowd at the average car show or race track, cams are a topic of considerable debate and often the supposed experts have very little hard data to support their various claims. Reasons for this are numerous, but among professional builders and racers, a cam-and-engine package that really performs better than the competition is a competitive advantage, and therefore it is a closely guarded secret. The enthusiast who only builds an engine every few years simply does not have the resources to make comparative tests, in which everything else is held equal, and won't have the same level of knowledge as a professional engine builder.

The main thing to learn from the following discussion is this: If the basic selection is reasonably close to the ideal cam specs, the differences between the cams in the plethora of appropriate choices will be modest. Hence, there is no "right or wrong" cam for a given application. Therefore, within a rational range, one cam will deliver one performance characteristic better than another, but on the flip side of the coin, certain other characteristics are not as good as with other camshafts. In other words, one camshaft can't be everything to every engine. You need to decide which characteristic is most important to you before you choose. To help you make that choice, guidelines for selecting a camshaft are provided later in this chapter.

Camshaft selection is more about context within the build than about any one set of parameters. The *combination* of parts defines the outcome—not any single item. Getting the absolute right set of cam specs is critical in a race engine, but much less so in a street car because driving behavior is more important than finding the last 10 hp. Don't become fixated on a particular specification or dimension. There is a reason that professional race teams have dozens of cams in stock; even with all their resources, they still need to try them out to find that elusive "perfect" part.

This FE hydraulic-roller cam with an induction-hardened core as indicated by the heat-induced coloration. These cams combine common hydraulic-lifter features with the roller-wheel design. These cams offer reduced frictional losses and cam manufacturers offer aggressive cam profiles in hydraulic-roller designs.

FE Camshaft Design Specifics

The Ford FE engine does have a few unique design features that need to be considered when selecting a camshaft. Most of these are well addressed by the various camshaft manufacturers and need only a cursory inspection when preparing to install your chosen piece. But they do need to be checked.

The FE uses a single dowel pin in the front of the cam as a locator and drive item. This pin has to be a snug-type (not press) fit, and it should lightly bottom out in the drilled hole in the cam. The original Ford assembly had the pin press fit into the timing sprocket, but the aftermarket timing sets have long since abandoned that practice. On one-piece fuel-pump eccentrics, the pin must be long enough to extend through the timing sprocket and come flush to the face of the eccentric. On two-piece pump eccentrics, a small tab extends rearward into the sprocket, and the pin must be recessed enough from the sprocket's face to accommodate the tab.

You sometimes need to make a small spacer to slip in behind the pin when it's not long enough. Other times you end up shortening the pin. This needs to be measured and corrected before installing the pin into the cam because it can be a real challenge to remove without damaging it.

The front cam bolt hole on an FE is a 7/16-14 thread from Ford. Most aftermarket cams are the same, but I have found some with a 7/16-20 fine thread tapping instead. The difference in thread pitch can damage threads if left unchecked. And I often find that the tapped hole is rough and full of debris. It seems that the production cleaning process misses that area regularly. The washer behind that cam bolt head needs to be very thick and strong. Hardware-store-quality parts do not work. On one-piece fuel pump eccentric assemblies, the washer must be large enough in diameter to overlap and retain the dowel pin. If you are running without a fuel pump eccentric, the washer bridges the opening in the sprocket and a thin washer bends and deforms.

The FE camshaft has all five bearing journals at the same nominal 2.124-inch shaft diameter. The number-2 and -4 journals need a groove in their centers for a 427 side-oiler block. These grooves route oil from the valvetrain transfers through to the deck. Enlarging that groove flows more oil to the heads, but that is rarely necessary. The grooves are not required with the common blocks such as a 390 or 428. Those

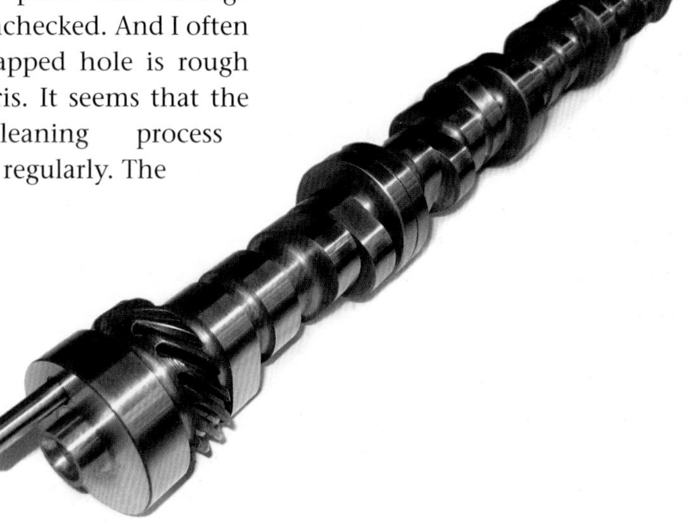

Solid-roller cam has a copper plate on a billet core. The copper keeps the center portion of the core softer and more flexible. Solid rollers deliver the best performance for most high-horsepower applications and race engines. The design provides virtually unlimited RPM potential, and it tolerates (and requires) very high valvespring pressures. But these cams require more maintenance than a hydraulic flat tappet and often are not the best choice for a street engine.

blocks have annular grooves already machined into the bores where the cam bearings are pressed.

The distributor and oil-pump drive gear is behind the front journal. I invest some time checking the teeth for burrs and rough edges. A few minutes spent with a wire wheel, a small file and a knife sharpening stone help ensure longer gear life, especially useful when running a bronze gear on a roller cam application.

The front face of the cam runs against the thrust plate. The cam sprocket slips onto the small snout that extends out from the front. Both of these surfaces need to be clean, flat, and free of any burrs or raised spots.

Cam Thrust Plate and End Play

A cast-iron thrust plate that is surface ground on both sides retains the FE camshaft. A pair of 7/16-14 fasteners secures this plate to the block. On original engines, these fasteners are Phillips-head or a button-head, hex-drive-type screws. A normal hex-head fastener usually clears the aftermarket double-roller timing sets, but you should check to be certain.

The thrust plate, sandwiched between the cam sprocket and the camshaft, allows about .005 inch in cam end play, which should be measured and verified. The plate can be polished to gain some clearance or the sprocket altered to reduce it. Original thrust plates can often be reused, but they do get damaged and worn from debris.

Blue Thunder typically offers replacement thrust plates in bronze, and these are an optional upgrade for enhanced compatibility with the steel alloys found in roller cams. Roller-bearing thrust assemblies are also available, and are included with the Danny Bee timing belt drive system. I've had good luck simply running the factory parts in most applications.

Camshaft Bearings

Cam bearings for FE engines are babbit lined and are manufactured by either Durabond or Federal-Mogul.

Standard center-oiler FE blocks all use the same cam bearing set. The inside diameters are identical, but outside diameters get smaller as you go farther back into the block. This allows automated installation in the factory, so that one mandrel can be loaded with all five bearings and pressed into place. Side oiler blocks use a different drilling to feed oil to the rocker arm assemblies, but share the same dimensions with their more common center-oiler brethren.

Aftermarket blocks use a variation on the side-oiler cam-bearing design. As with the factory parts, all the journals have identical internal dimensions. But, in aftermarket blocks, the outside diameters are the same for all five bearings. The Genesis blocks use the same diameter as the factory number-1 bearing

A cast and surface-ground cam thrust plate is used on all FE engines, and it's the same as on most other Ford engines. These plates provide essential support and absorb the cam load when the engine is in operation.

This one-piece fuel-pump eccentric is a Ford Motorsports replacement for a 460, and it works fine for use on an FE. The listed application is for a 460 with a 7/16-14 fastener.

for all positions, but have Ford-style side-oiler drillings. The Pond block uses the same strategy, but with slightly wider cam bearings throughout for greater load capacity. These require trimming the front cam bearing for distributor clearance after installation.

Federal-Mogul part number 1268M is the bearing set designated for Genesis blocks, and I have used it as a repair part for servicing damaged factory blocks when a cam bearing has been spun in its bore.

Roller cam bearings have been occasionally used in Super Stock level engine builds. The aftermarket blocks have additional material in the cam tunnel, making installation easier. The parts designed for 460-style engines can be made to work. The benefits of the roller bearing conversion are small, difficult to quantify, and often disputed, but it is a viable repair to a damaged block that may otherwise be discarded.

The standard FE hydraulic flat-tappet lifter features a hairpin-style retainer clip. These are quiet and reliable lifters that came as original equipment in the vast majority of FE engines. While these are suitable for most street-oriented engine builds, they are not preferable for high-rev racing applications.

Selection Criteria: Lifter Type

First you need to decide on the lifter style for the engine. There are four basic options: hydraulic flat tappet, solid flat tappet (also termed "mechanical"), hydraulic roller, and solid roller. Each variation has advantages in certain types of use.

Keep in mind the fact that flat-tappet lifters rotate against the cam by spinning in their respective bores; it is not a sliding contact. The bottom of a so-called flat lifter actually has a crowned profile that works in concert with a taper ground onto the cam lobe to promote the spin. You can see and check that radius by holding a pair of lifters against each other "foot to foot" in front of a light background. The profile will be readily visible. Important note: Lifters that are truly "flat" are worn out and should be replaced.

Roller lifters do not rotate in their bores—they actually have a bar linking two lifters together to prevent rotation. They use roller bearings that ride against the cam. The cam lobe is thus ground flat without a taper. The lifter's roller has a small radius ground into it to accommodate any *minor* variance in lifter-bore geometry.

Hydraulic Flat-Tappet Lifters

These lifters are the original equipment in the vast majority of FE engines. These inexpensive, quiet, and reliable lifters perform perfectly well in most street-oriented applications in which strong low- and mid-range performance with minimal maintenance is desired. The FE pieces share the same .874-inch diameter as other popular Ford lifters, but common Ford lifters do

An FE race-style hydraulic lifter has the spring-steel retainer rather than the hairpin-style retainer clip. The stronger clip keeps the plunger in place if valves float, and therefore are a far better design for high-rev applications.

not have an oil hole in the pushrod seat. This is because the traditional FE does not oil through the pushrods (the routing path is covered in detail in Chapter 5).

Hydraulic lifters operate with an inner plunger that floats on a cushion of oil within the outer shell. Oil is fed under pressure into the cavity separating the two sections and exits through the clearance between the inner and outer parts (or through the pushrod feed hole if so equipped). A check valve prevents oil from exiting through the feed orifice. The controlled leakage allows the lifter to run with a small amount of preload, which takes up any clearance in the valvetrain, and to compensate for wear. Most lifters will have roughly .100 inch of plunger travel between the upper retaining clip and the bottom of the cavity, with the desired operating position centered in that range.

Hydraulic lifters for the FE fall into two groups: the stock replacement

FE dumbbell solid lifters are unique to the engine, but small-block lifters work as long as pushrod length is correct. Solid lifters are a great choice for a budget-oriented high-power or high-RPM combination. They are subject to careful break-in procedures, but once running with the correct high-zinc lubricants they deliver excellent power and reliability. If you intend to oil the rockers through the pushrods with TD- or Jesel-style rockers you need to use the small-block-style lifters.

parts or the so-called "anti-pump-up" race styles. In reality, the only major difference between lifters is the use of a heavier-duty retaining clip on the performance parts. This clip allows the performance parts to operate at or near zero preload without the risk of the lifter coming apart. While these are stronger and safer than standard parts, they do not offer a dramatic performance advantage. Perhaps a couple hundred RPM can be gained from running at zero lash, but any wear or temperature-induced dimensional change results in noisy operation and the need for adjustment. Hydraulic lifters were installed to avoid those issues in the first place.

Hydraulic Roller Lifters

These lifters combine the low-maintenance features of the traditional hydraulic lifter with the roller-wheel design. These offer reduced wear, reduced frictional losses, and access to some enhanced cam profiles—plus break-in is not required. With the reduction in extreme-pressure additives in modern oils, notably less zinc and phosphates, the break-in period on flat-tappet cams has become a much larger issue than it was in the past. While successful break-in is certainly achievable without a lot of drama, the hydraulic roller retro-fit is an inexpensive way to completely bypass the issue.

Hydraulic rollers were never installed in any factory FE applications. The engine was discontinued long before these came into production. But the aftermarket has responded to the current demand by offering new parts that drop right into a normal FE block. The lifters use a link-bar design to prevent rotation, which is similar to solid-roller applications. All that is required are the lifters, cam, and custom-length (shorter) pushrods. Since FE hydraulic rollers are newer parts, they have the ability to oil through the pushrods if desired. A true FE roller lifter has the oil-feed hole perpendicular to the roller axle. A lifter intended for a 460 engine has the oil hole in-line with the axle, and puts way too much oil up to the rockers if used with the TD system.

The hydraulic roller lifters are large and heavy, and the only downside is an effective limit on RPM. Combined with the hydraulic design, they are best suited for moderate-RPM street engines with cams intended for a peak RPM at or below 6,200. In dyno testing, I've seen a pretty dramatic power drop off once the valvetrain control is lost on hydraulic-roller installations. Adding more spring pressure helps, but the hydraulic design can only accommodate so much pressure before becoming noisy and inconsistent in operation. There are ways to work around the limitations but unless race rules mandate the use of a hydraulic lifter, other types are better suited for those needs.

Solid Flat-Tappet Lifters

These lifters are the best choice when you are trying to make horsepower on a budget. They allow the 7,000-and-up-rpm levels that fall within the capability of the traditional FE block architecture. They are simple, inexpensive, and easy to install.

The solid flat tappets used in original FE engines have a "dumbbell"-shaped design, with a reduced diameter through the center of the lifter. There is no real advantage to these, but they're still considered the "normal" FE solid. If you are intending to oil the rockers through the pushrods for the TD- or Jesel-style rockers, you need to use common small-block Ford solid lifters.

The downsides to a solid lifter include the need for careful break-in and proper high-zinc oil, along with the requirement for periodic lash inspection and adjustment. In addition, a solid lifter accelerates the valve off the seat very quickly because of edge contact against the cam. However, the shape of the lobe limits its opening rate once in motion.

Solid Roller Lifters

These are the preferred choice for serious high-horsepower applications.

The basic design allows virtually unlimited RPM potential, and it tolerates (and requires) very high valvespring pressures. The cam profiles for solid rollers can be extremely aggressive with very high valve lifts—the roller can accommodate radical lobe designs. As the diameter of the lifter's roller increases, the cam can become even more radical with improvements in both cam-to-lifter geometry and durability due to the larger parts.

The standard FE lifter diameter is the same as other Ford engines at .874 inch, but I have gone to .904-inch diameters in serious race applications. Larger-diameter lifters offer advantages in terms of strength, and on flat-tappet cams they allow more aggressive lobe profiles. You can use bronze lifter bore bushings to correct for any variances in lifter-bore geometry, although I've had good success just running against the block's cast-iron bores. I see far more lifter-bore-related issues in older blocks from wear, moisture, and rust than I do in new replacement blocks.

The link bar is the most common roller-lifter design used in FE applications although other options do exist for high-end race engines. In order to send oil through the pushrods, you need to use appropriate FE lifters. The oil-feed hole on the FE lifter is at a 90-degree angle to the roller axle. Lifters from a 460 do physically fit, but the oil-feed hole is parallel to the roller axle, and sends far too much oil to the valvetrain when using pushrod oiling.

FE solid roller lifters accommodate high spring pressures, and therefore are the lifter of choice for racing and serious high-performance engines. This set of Comp Cams lifters shows the proper FE oil hole orientation, which is toward the intake valley.

Cam Specs: Lift, Duration and Lobe Separation Angle

Cam specifications are given in valve movement increments. The various camshaft manufacturers provide the lift as well as duration specs, and most builders select a camshaft based on those numbers. The manufacturers provide gross valve lift, an advertised duration number, durations at .050-inch lift, and the lobe separation angle. This is good and useful information, but only a small part of camshaft lobe design. Professional cam designers work with lobe acceleration rates and opening/closing events to arrive at the desired valve motion. The duration data and lobe separation points are thus an output from the effort, rather than an input. Therefore, while directional information can be derived from those specifications, you cannot assume that they are the entire story. Context is the key in comparing similar cams.

Lift

Lift is the maximum distance that the valve is moved from its seat, measured in thousandths of an inch. Gross valve lift, in increments of 1 inch, is determined by multiplying the cam lobe lift by the rocker ratio. This definition includes a lot of assumptions. First, the rocker ratio must be accurate, which is rarely the case. Next, this does not reflect the impact of valve lash. Last, the valvetrain always has some flex or deflection. It is not unusual to see considerably less actual lift when it is measured at the valve. Checking lift at each spot in the valvetrain tells you where the loss occurs by comparing your actual measurements to the mathematical ideals. Reducing deflection through use of stronger rockers, heavier-duty pushrods, and better mountings yields a significant improvement in both performance and durability.

If your valve lift is beyond the best flow range of your heads in a race engine, that is acceptable; but the port should not become turbulent or "back up" at the higher lift. This is because of the next variable, which is the amount of time that the valve can be held open at that peak level for maximum flow through the port.

Duration

Duration is the amount of time that the valve is open, as measured in crankshaft degrees. Since the crankshaft turns twice as fast as the camshaft, the number of degrees quoted is comparatively large, as 720 crankshaft degrees occur per one full turn of the cam. Advertised duration numbers are of nominal value because there is no real standard for the measurement—some cam companies use .002-inch lift for this measurement, others use .006-inch, others a "zero" base. The "duration at .050 lift" specification came about as a method of industry standardization—a way to compare cams from different suppliers. In the crudest of simplifications, the duration determines the operating RPM range of a camshaft.

Duration numbers (both advertised and at .050-inch lift) are useful for guidance but not definitive when comparing cam profiles. Various software programs can get you "close" on basic cam selection using these values, but plotting the full range of valve motion throughout the lift curve is the only accurate way to compare cam lobes. You can easily find two cam lobes with comparable lift, advertised, and "at .050" numbers, but they are significantly different at every other point in the range of motion. Comparing cams would be better done with a graphed curve showing durations at, for example, .100- .200-, .300-, .400-inch, etc. None of the large cam companies currently provide that level of detail for consumers.

Lobe-Separation Angle

This refers to the difference between the intake- and exhaust-lobe centerlines as expressed in degrees. The basic assumption is that cams having a wide angle of 112 or more degrees are street oriented for a broad powerband and smoother idle, and that separation angles of 106 to 108 are race-car material with stronger peaks and a narrower power band. These ideas contain a grain of truth because a cam designer may use the same lobes and just move them around to change the tuning behavior. But they may also use completely different lobes to get the performance characteristics they desire, changing the angle values dramatically without changing the tuning behavior.

Intake-Lobe Centerline

This is expressed as a number of degrees before crankshaft top dead center. This specification is used to install the cam at the proper position relative to the crankshaft, as desired by the cam designer. Cam lobes are not symmetrical, so the centerline is not going to be at the point of maximum lift. Instead, you find the centerline by locating a point .050 inch lower than peak lift on the approach and descending sides of the lobe, and splitting the difference. Installing the cam at the centerline position specified by the manufacturer should result in the opening and closing events occurring per specification. When cross-checking the opening and closing events, you must use the proper lifter. A solid lifter gives bad data if used for checking on a roller cam. The lobe's flanks are completely different, even though the event specs may appear to be similar.

Advancing or retarding the cam changes the opening and closing events relative to the crankshaft, and alters the installed intake lobe centerline. It does not change the lobe-separation angle, lift, or duration values because those are ground into the cam itself. Altering the installed position by a few degrees can have either a very modest or a very significant impact on the way the engine runs. This is a tuning change with results that cannot be easily predicted, and therefore, you have to try it.

So How Do I Pick a Cam?

There are a few ways to go about the cam selection process. You can open a catalog and find the one that most closely matches your desires based on the manufacturer's descriptive information. You can call the cam company's tech lines and get more detailed recommendations. You can consult your friends, engine builder, or machinist to benefit from their experiences. Or you can recruit a professional cam designer to create a cam profile matched to your specific needs.

None of these are wrong, but some methods are going to have a better chance of success. The engine builder, cam designer, and tech department are all likely to have considerable experience with the kind of questions that arise. In the case of the designer, he or she works from your combination to generate a profile, while the tech group tries to match a pre-existing catalog profile to your application. Professional racing efforts almost exclusively use designer-specified solid-roller cams. The designer likely ends up with the stronger combination, but that increment of power comes with a cost that is potentially out of context for a more modest street-type build.

I'll provide some general guidance as to what has worked for me in various FE engine builds. As noted earlier, this is not to be taken as definitive/prescriptive—but as general guidelines. Your results will certainly vary depending on the engine characteristics, including displacement and compression, as well as car variables, such as transmission, rear-end gearing, and your own perceptions and expectations regarding drivability, sound, and power.

One statement that always holds true is that a larger-displacement engine can use a longer-duration cam with good results. Since most FE cam catalog recommendations are based on standard displacements, the now-popular stroker engines can comfortably go a step up in cam sizing without ill effect.

On 390-based 445-ci stroker engines, I often use the Comp Cams 282S solid flat-tappet cam. With 236 degrees of duration at .050-inch lift, it delivers a noticeable but not too choppy idle at around 750 rpm. Gross valve lift is noted as .571-inch; subtract .022 lash to get a net lift of .549 inch. This is low enough to use the springs included with the popular Edelbrock heads, although I usually change them out for a double spring. Power peaks at about 475 hp at 5,500 rpm, with 503 ft-lbs of torque.

You will get more power and a higher RPM peak of around 6,000 by going to the Comp Cams 294S grind, with 248 degrees of duration at .050-inch lift and .605-inch gross lift. With about 20 more horsepower and more torque throughout the power range, this seems like a winning package, and it is. But it has a much choppier idle at 850 to 900 rpm and reduced low-RPM throttle response.

Putting a .668 gross lift, 248/252 duration at .050 solid-roller cam into a similar 445-inch engine got beyond the normal working range of the rest of the budget-oriented combination and only netted an extra 20 peak horsepower, while costing hundreds of dollars more. Of course the idle sound would stop you in your tracks from across a crowded parking lot, if that is one of your desired outcomes.

Now go to a 482-inch 427-based stroker and the situation changes dramatically. That same 236 at .050 282S cam we started with now delivered 537 hp at a similar 5,700 rpm. The torque also moves up to 574 ft-lbs at 4,200 rpm and idle is nearly smooth at 700 rpm. It also delivers excellent driving characteristics in a much tamer-sounding package until you hit the throttle and the torque takes over. But it runs out of steam pretty early for such a racy short block, making this a great engine for a Galaxie cruiser or a truck.

More common with large displacement engines are solid rollers. I've run the same .696-inch lift, 258/264 duration at .050 solid roller in multiple combinations with interesting and dramatically different results. With an 11:1-compression 482 and ported Edelbrock heads, this cam is good for a bit over 610 hp at 5,800 rpm, along with 620 ft-lbs of torque. Put that same cam into a 520-inch, 12.5:1, dual-quad, CNC-ported high-riser-head combination and you get up to 771 hp at 6,900 rpm, along with torque of 653 on the same dyno.

It seems counter-intuitive that the same exact cam would peak at a much higher RPM in the larger engine. What these results teach us is that the *combination* is the important item—not the individual components. The cam was not the limiting factor in the 610-hp engine; the heads were.

Working with the common catalog cam profiles, I can make some basic recommendations. These are dramatic simplifications, but provide a starting point. On a 390 engine, keep the duration at .050 down to a range between 215 and 225 degrees for a mild idle, and 230 to 240 degrees for a choppy one. As you go up in displacement, you can also go up a similar amount in duration and retain comparable driving and idle characteristics. An increase of 10+ percent in displacement to a 445 can easily handle a cam with 235 degrees of duration at .050 lift with very good street manners. It's something of a non-linear sliding scale, though, and a 482-ci engine only wants around 250 degrees for a comparable performance range. A cam with durations of 260 degrees at .050-inch can support a 482-ci engine beyond 6,500 rpm, assuming the rest of the parts are strong enough for that operating RPM.

For lift, I try to keep the milder street combinations at or below the .600-inch-lift range. The reasons for this are twofold. First, the Edelbrock heads are delivered with a spring and retainer package oriented around a .600-inch peak lift. Second, keeping the lift within this window reduces likelihood for geometry issues or pushrod-to-intake interference.

On more serious street/race engines, I push this to about .700-inch lift. This gets into an area where the manifold and valvetrain must be checked and corrected, but still delivers good durability once everything is properly set up. I use an oval-track-oriented valvespring and a tighter lash setting on roller-cam builds in an effort to get good component life. My feeling is that excessive loose lash and inadequate spring pressures are what really "kill" roller lifters in street use.

Going back to our original discussion on selection, there are no perfect, right answers. Engines cannot read. Use the information in this book and the most experience you can acquire and afford, and then make an educated camshaft choice. Then you're going to have to go and try it in your car. Only then will you know whether it works to your satisfaction.

Timing Systems

With the basic cam selection information defined, let's move forward with the next piece of the puzzle—the timing set. FE engines use a timing assembly very similar to other Ford engines, with the aforementioned cam thrust plate, an offset positioned dowel pin for locating, and a single central cam bolt for retention. The factory cam sprockets were often aluminum with molded nylon teeth for quieter operation. The chains were a link type with single-toothed sprockets. The steel lower sprocket has a light press fit onto the snout of the crankshaft and is located by a single keyway.

The original cam drive assembly also included a "C"-shaped spacer washer sandwiched between the upper sprocket and the cam itself. All current replacement timing sets incorporate this spacer into the sprocket; the spacer, if found, should be discarded during service. An oil slinger goes in front of the sprocket to reduce the amount of spray aimed toward the front seal.

Cloyes Tru-Roller timing sets include pre-stretched timing chains from Renold or JWis. Sprockets are accurately made, and therefore the ideal length is maintained. In my experience these have proven to be simple and reliable.

Factory Timing Sets

Link-type timing chains are comprised of multiple side plates woven together and assembled into a chain with pressed and peened pins. The side plates themselves are in contact with the single wide teeth on each sprocket; the pins do not touch the teeth. The design is simple and quite durable, but has fallen out of favor in the performance market over the past twenty years. You won't find many of them in performance applications, but most stock replacement parts are of this design.

The early nylon-toothed sprockets are the likely culprit for the "bad rap." The nylon would become harder and less flexible over time. That tendency, combined with the eventual loosening of the chain due to stretch and wear, caused the sprockets to shed their teeth, usually on cold startup. Replacement sets use sprockets machined completely from iron with no plastic and thus eliminate this potential problem. Another issue faced by link-style chain systems is that the links are in contact with, and eventually wear into the sprocket teeth.

Another weakness in link-style timing sets has nothing to do with the design at all; it's more about market economics. Since they are viewed as stock replacement items, these timing sets are very low cost and therefore quality has suffered over the years. Unfortunately, imported components of questionable quality and corner cutting in both quality and manufacturing are common.

Roller Timing Systems

Roller timing sets are the norm in high-performance applications. Every roller set for the FE engine is a "double roller" design, having two identical rows of narrow teeth on each sprocket. The roller chain has only one pair of side links per segment and is driven through the connecting pins.

Roller chains can be of either an "ANSI" (American National Standards Institute) design, which has the sprocket teeth riding directly against the pins themselves, or of a "BSI" design where there are freely rotating bushings surrounding each pin. The ANSI chains are less expensive, and are sold with pins in .200- or .222-inch diameters. The "BSI"-type chains have a pin bushing diameter of .250 inch. The latter design is considered superior, and is referred to as the "Tru-Roller" series in the Cloyes literature. It is important to identify which chain you are using because the sprockets cannot be interchanged; the tooth profiles are different.

Upon initial use, all chains stretch to an extent because of construction and function of the part. Punch-formed holes in the links have something of a "point contact" until a smooth bearing surface is established. Most quality chains are "pre-stretched"; they are mounted over a set of rollers and pulled under load to a specified length.

Brand Names

With the notable exception of Cloyes, timing sets are generally outsourced items—sold under a marketing brand name that is often not related to the actual manufacturing source. Chains and sprockets are manufactured by numerous companies in various countries, and in a wide range of quality.

Chains can usually be identified by miniscule stampings on the links. Manufacturers of high-quality performance chains include Morse from the United States and Mexico, Renold from France, JwiS from Germany, and Daido (DiD) from Japan. Rolon from India is a common supplier for lower-cost sets.

Sprockets can be harder to identify. Cloyes makes its own, as well as private-label parts for numerous other companies. SA Gear provides many sprockets to the various marketing companies, but does not sell to consumers. There are numerous offshore sprocket suppliers as well, with quality levels ranging from excellent-appearing billet pieces to, well, just plain junk.

With such a jumbled and confusing market situation, the only thing you can do is to pick a brand based on reputation. I have gone to using exclusively Cloyes Tru-Roller timing sets. The reasons are twofold.

First is that I know who makes them and the parts are always consistent from box to box. There are no mixed parts from multiple suppliers that change with the vagaries of international currency fluctuations. Second is that the basic Tru-Roller sets simply flat-out work—no fancy quasi-billet stuff. There are only three keyways, but they always seem to fit right and deliver the timing position I'm after.

Cloyes offers Tru-Roller sets with reduced center-to-center distances of .010 and .005 inch. These are intended to address issues caused by excessive line boring but I've rarely had to use them. On several occasions, customers have requested the reduced center-to-center sets from other manufact-urers to correct a perceived loose-chain problem. However, the standard Cloyes timing chains and parts remedied the problem.

Billet, 9 Keyway and Thrust Bearings

You can purchase some really cool-sounding timing sets these days for your FE project. There are sets with nine keyways instead of only three. There are sets with billet-cam sprockets. There are sets with thrust washers and thrust bearings. In my experience, most of these innovations don't offer much of an advantage, considering the added cost and complexity.

While the nine-keyway deal sounds like a great idea to really dial in your cam timing, it makes a couple assumptions. The first is that the key slots are all labeled according to their actual function. Having tried a number of these over the years, it seems that they are usually labeled at random, so you wear out your wrenches trying to find the one that gets you what you desire. The second assumption is that you are going to notice whether the cam is installed at the perfect spot or if it's a degree off. The truth hurts a bit—nobody other than a NASCAR team or a Pro Stock builder will be able to document whether any position is "right or wrong" until they've run the engine with multiple settings. While we all strive for perfection, the ideal for cam timing is to get a

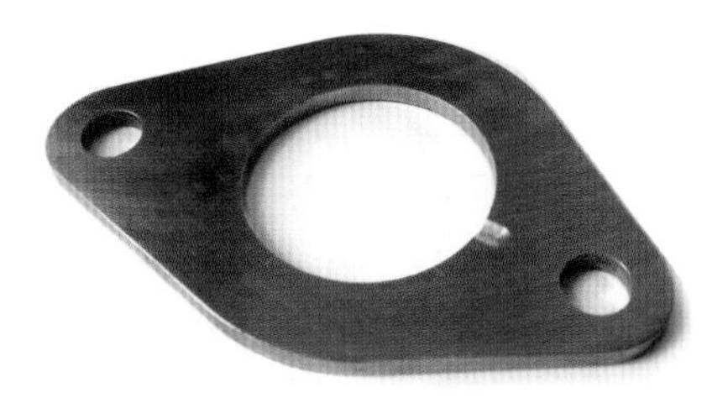

This bronze thrust plate for roller-cam applications is a Blue Thunder part—nice, but not mandatory. The simple Ford thrust plate seems to work just fine. If you're concerned with steel cam compatibility, this is a good option.

good-fitting system that is reliable, predictable, and consistent.

Billet sprockets look great hanging on the wall in a blister package. But beyond that they have no real functional advantage over a properly made hardened cast or machined steel sprocket. In 40 years, I have yet to see a double-roller cam sprocket that caused a failure, so the material has proven to be pretty good for the task. The machines that form the teeth and center machining are identical, so there is no dimensional advantage. All crank sprockets have (or should have) hardened teeth on a steel base.

Cloyes sells its infinitely adjustable "Hex-A-Just" timing set for the FE. As fond as I am of the traditional Cloyes set, the Hex-A-Just is not one of my favorites because it uses the fuel-pump eccentric as an adjuster, and pivots around the center bolt. The design counts on the center fastener for tightening and holding the cam position. The dowel pin is allowed to move in a slot in the sprocket and no longer serves as the locating device. Many folks use this system with good results, but I'm just not comfortable with it.

The thrust-washer and thrust-bearing kits strike me as a solution looking for a problem. Originally offered as a repair part for small-block Chevrolet engines, cam thrust washers can be used to fix a block with a worn thrust surface. This was never a problem on Ford engines, which used a thrust plate to handle the cam loads; but if you have debris induced wear, just replace the plate. Thrust bearings, using Torrington-style flat rollers, are theoretically useful in a flat-tappet-cam application with the cam-lobe-angle-induced rearward loads. Most race engines have a roller cam and, thus, no-angle-induced loading. The simple Ford thrust plate seems to work just fine. Blue Thunder sells a bronze thrust plate if you're concerned with steel cam compatibility.

Gear Drives

Gear-to-gear-type cam drives have largely fallen out of favor over the past 20 years because they offer no real performance advantages. But some folks really like the sound of the blower-like "whine" that they produce. As far as I know, the only supplier for an FE gear drive is Milodon. It is a system that fits under the stock timing cover and requires that a couple additional holes be drilled into the front of the block. If you plan to run one, you should acquire the system and do the drilling and tapping before block assembly.

Belt-Drive Systems

The belt drive is considered the best part to use in a drag-racing environment. It is relatively easy to install, provides reliable and consistent performance, and unmatched tuning capability. While performance advantages are sometimes touted by various belt-drive suppliers, they would be incremental at best, and largely offset by the added drag from the required camshaft seal.

The great advantage of a belt-drive system is that it allows cam-timing changes in a couple of minutes, with the engine completely assembled. The billet timing cover supplied with an FE belt drive is sealed, while the sprockets and belt are on the outside of the engine. With a belt drive on your engine, in the dyno cell, you can optimize cam timing in a few quick pulls. Cam timing is adjusted as follows. Loosen the half-dozen fasteners around the perimeter of the cam sprocket and align the appropriate marks while "bumping" the crank around with a wrench. It takes five minutes, tops. Those same tests and changes would take hours with a factory-style chain system.

Danny Bee produces the only FE belt-drive system. It's expensive but it makes cam-timing changes and swapping far easier.

Here is something you want to avoid. These are the wrong cam washers. The large-diameter washer is too thin and will deflect and fail. The smaller-diameter washer is thick enough, but will not retain the dowel pin. The correct washer size is at least .100 inch thick and overlaps the dowel.

In addition, the belt-drive system allows for much easier cam changes. Only the cam sprocket and center section of the front cover need to be removed; the damper and the cover itself can remain in place.

The only supplier for belt-drive systems for the FE is Danny Bee. While expensive at nearly $1,000, the system is of very high quality, and includes everything necessary to make the conversion. Instructions are a bit sketchy but you can use the ones for the more popular 460 Ford kit. The only thing to look out for is the lower sprocket, which may need to be shortened to provide proper belt alignment on certain factory crankshafts—notably the 391 truck forging.

Timing Covers, Damper Spacers and Pointers

There are a few different factory timing covers, but they effectively interchange with one another. All FE timing covers mount the same way and include a spot for the timing pointer and the front crankshaft seal. They have a flat surface at the bottom with 5/16-18 threaded holes for four of the oil-pan bolts, and a side opening with a pair of 3/8-16 threaded holes to mount the mechanical fuel pump. Four 5/16-inch fasteners secure the cover to the block around the upper circle, and four 3/8-inch fasteners secure the lower half. Several of the mounting fasteners extend into either water or oil—I recommend using a dab of sealant on assembly.

The earliest FE engines used a sheetmetal cover—an item not often seen on performance engines. The vast majority of FEs have a die-cast aluminum timing cover. The earlier version has two small fasteners retaining the timing pointer. Later ones use a small fastener for one end of the pointer and have an extended portion of the casting to allow a longer 3/8-inch bolt to hold the other end.

Most FE dampers have the timing marks engraved on them, thus the pointers have a simple angled end. The two pointers are functionally interchangeable, providing the appropriate cover is used. But some 427 engines only had a scribed line on the damper, with the coordinating pointer having a range of timing positions; these pointers are unique and valuable.

The front seal on an FE is pressed into the timing cover from the rear. It seals against the damper spacer, a cylindrical sleeve that slides over the crankshaft before the damper itself is pressed into place. The damper spacer is located by the same keyway that located the crankshaft timing sprocket. It is pretty common for the damper spacer to develop a worn-in ridge where the seal rides, causing oil leakage. Slide-on repair sleeves are available, as are replacement damper

This is a two-piece fuel-pump eccentric. The inner bolts hold the sprocket and cam while the pump itself rides on the outer ring. On two-piece pump eccentrics, a small tab extends rearward into the sprocket, and the dowel pin needs to be recessed from the sprocket's face enough to accommodate the tab.

This is the correct dowel-pin location for use with a one-piece fuel pump eccentric. The pin needs to be long enough so it extends through the timing sprocket and fits flush to the face of the fuel-pump eccentric.

spacers, in both aluminum and steel. If installing a repair sleeve, remember to lubricate the seal surface. Centering the damper spacer in the seal before bolting down the timing cover can help prevent premature seal wear and subsequent leakage.

Dampers

The damper used on factory FE engines is of conventional design—an inner hub, a bonded elastomer layer, and an outer ring. Some early dampers had a pulley groove machined into the outer ring, but most have pulleys bolted to the front hub section. As noted above, most FE dampers have the timing marks engraved on their outer ring. FE dampers are located by a single 1/4-inch keyway in the crank.

Although a large array of factory parts is available, their relative merits and part numbers are best kept within the context of a restoration-type build. Most of these are now approaching or exceeding 40 years in age. The rubber has long since surpassed its service life and will show signs of cracking and wear. With aftermarket replacements being readily available and inexpensive, there is no reason to reuse an original part unless you are doing a restoration or are on a restrictive budget.

OEM Ford FE timing covers are easily installed and include a spot for the timing pointer and the front crankshaft seal. Early timing covers have two small fasteners for the timing pointer.

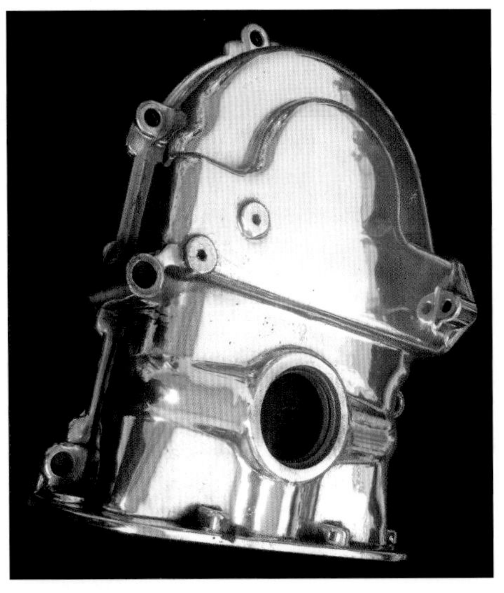

Later timing cover with multiple timing pointer bolt locations. Usually the larger hole is used along with one small one.

Genesis sand-cast timing covers are functional, but heavier than stock parts.

The least-expensive offshore dampers from Professional Products are suitable for street use, and are likely better than a worn-out factory piece. They have timing marks around the perimeter and come with a universal pointer that fits the common timing cover.

Romac products offer the next step up in quality. An Australian company, Romac supplies a very nice-looking reproduction of the factory 427 damper, as well as a top-quality performance damper with an aluminum outer ring. Although cosmetically pleasing, its silver color is hard to read with a timing light. Both are SFI certified, making them legal for use in NHRA drag-racing venues.

The best dampers are likely from ATI. As a supplier to the majority of professional race teams, the quality of ATI dampers is well established. The only issue, beyond the higher but appropriate price, is that its FE damper is a variation on other parts. It has an outer ring from a 460, which means that the timing marks are off by a few degrees, but timing needs to be checked anyway in a race engine. Also, the hub has the big-block Chevrolet bolt pattern; thus, FE-specific pulleys require modification.

Damper mounting bolts on FE engines are large at 5/8 inch and rarely strip out. ARP sells replacement bolts with the required big washer. The bolt on a 391 truck crank is larger yet (3/4 inch), but ARP sells a bolt for the KB hemi that works just fine. These must be tightened to specs and used with red Loctite for final assembly, or they will come loose.

CHAPTER 10

INTAKE MANIFOLDS

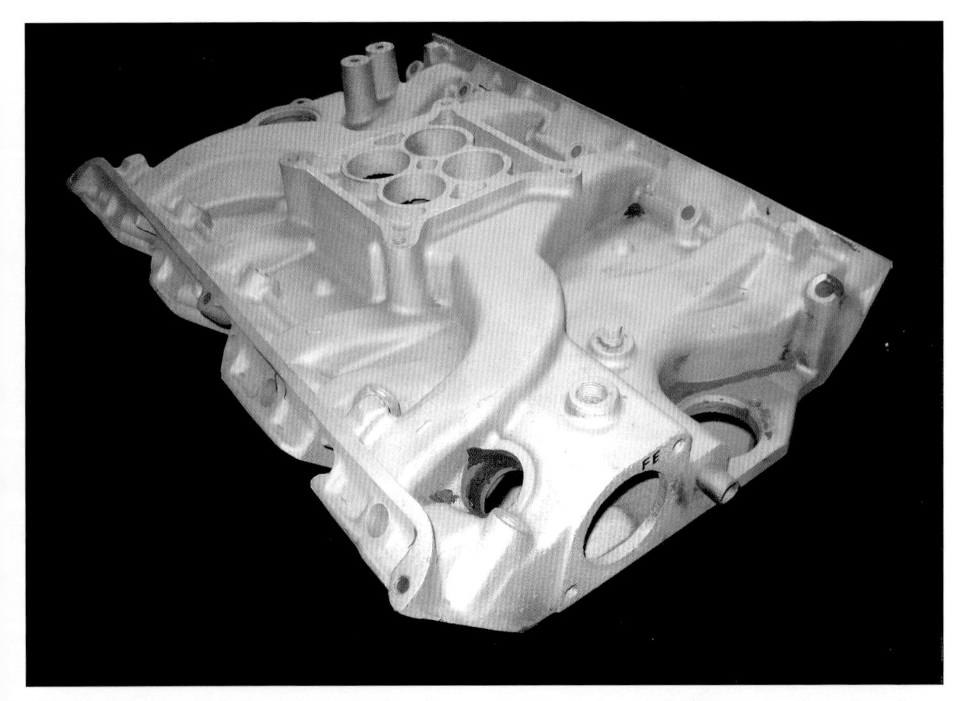

FE dual-plane intake has over/under runners and an oil filler opening. Dual-plane designs such as this are ideally suited for high-performance street use because they deliver quick throttle response, superior acceleration, and provide good flow characteristics at a wide range of RPM.

Ford FE intake manifolds are unique to this engine family, and instantly recognizable when compared to any other engine's parts. One third of the valve cover extends over the intake casting, and the pushrods run through cast or machined passages in the manifold. The FE manifolds are comparatively wide and heavy, and have numerous machined surfaces that interface with the heads, block, distributor, valvetrain, and valve covers.

With an original manufacturing and performance use history that extends over a 50-year period, virtually every intake fashion and style has been produced for the FE, albeit in far fewer variations than for the more popular engines. The builder can choose from original or aftermarket; single 4-barrel, dual-quad, three 2-barrels, six 2-barrels; tunnel rams or cross rams; single plane or dual plane; aluminum or iron. Depending on the desired cosmetics and performance any of these choices can be the right one for a given project.

Rather than trying to detail every available original part number or aftermarket option, or selecting a "winner" among the plethora of intakes available, I'll just cover general guidelines and the most popular choices. If you are interested in historical research (the FE has a vast amount of outstanding racing history), there are other publications that better serve those needs.

Jay Brown, a well-known FE enthusiast and racer, has done exhaustive dyno testing on literally dozens of FE intake manifolds. His efforts have brought some unexpected "winners" to light, and quashed a few urban legends as well. The dyno testing does a great job of delivering comparative wide-open-throttle (WOT) performance but does not quantify part-throttle behavior. Some of the commentary found in this chapter is based at least in part on Jay's efforts.

Intake Interchangeability

As a lead in to the manifold discussion, it is important to note that every intake does not physically fit onto every FE engine. But the vast majority of them do. The exceptions are the same ones noted in Chapter 6.

The 427 SOHC obviously uses a unique intake that does not at all resemble any other FE part. Closer to a 429/460 intake from an external appearance, the SOHC intake does not have the characteristic FE valve cover rail. Relatively few variations of intakes are or have been available for Cammers, such as single and dual 4-barrel units, either reproductions or originals from the factory, and a few single-plane iterations from small specialty suppliers such as Dove. You won't find one in the local swap meet or yard sale.

The 427 tunnel-port intakes use a unique bolt pattern and port layout; they do not mount to any of the more common FE wedge heads. At a glance, these look like normal FE parts, until you see the huge round ports. As a rare and valuable specialty item, a tunnel-port intake is not likely to be in the hands of an unknowing seller. Dual-plane and single-plane variations were produced, but always in very limited numbers and never in a production vehicle.

The 427 high-riser intakes are the "tricky" ones. They are somewhat more common than the tunnel port, and aftermarket ones are comparatively regular finds in Ford swap meets. They look like they fit on your normal 390 or 428 with a little work, but they don't. The valve covers, and hence the gasket surface, are in an altered position and on a different angle than so-called "normal" FE engines. The port flange is considerably taller. So if you find a high-riser intake—even a really cool looking dual quad or injector one—you're best off letting it go unless you have or are willing to acquire the matching set of heads.

Every other stock FE intake manifold physically fits on every other FE engine that was not mentioned above, although some combinations may require some machining and other work. As noted in Chapter 6, medium-riser and low-riser heads are best defined by their comparative port floor position. The same is true of intakes. The port roof and sides are reasonably close to each other in dimension and well within normal port matching range. Although mismatched floors on port openings are not commonly considered desirable, such packages do certainly run. And sometimes, they run surprisingly well. The mounting bolt patterns are all the same, as are distributor openings and coolant passages.

Recommendations First–Then the Details

There is no single "right" answer when choosing an intake manifold for your FE project. You need to select an intake manifold based on a number of factors such as cosmetics, operating RPM, vehicle weight, transmission type, and engine size. Your selection will reflect a combination of vehicle components and your performance goals. Here are a few quick recommendations on those parts that are still available new.

If you are building a street-oriented FE using the popular Edelbrock heads, you would be well served to go with the coordinating Edelbrock Performer RPM intake. Good power and a good price point.

If you are building an engine for a "restomod" or an original-looking package like a Cobra, consider the Blue Thunder dual plane. It just looks "right." It's also the choice for an earlier car because it has the needed oil-filler tube. This is also the best pick for shaker-hood-equipped Mustangs because the Edelbrock intake's carb pad is in the wrong spot.

When mounted in the common fashion, the front bowl of the carb hits the distributor, and therefore the FE 2x4 carbs are mounted backward.

If your FE is running dual quads, I recommend the Blue Thunder dual-plane or the Dove Tunnel Wedge single-plane. Both have a factory appearance, with the Blue Thunder being restoration-style muscle and the Dove being a bit more race-car oriented. Both look proper on a vintage ride with the finned oval air cleaner, fuel log, and linkage.

These selections are not intended to be limiting or to be perceived as the only correct choices. They are just simple, proven picks from parts that are readily available. There are lots of other aftermarket and factory-original options out there.

Original Parts

Factory Ford intake manifolds are almost exclusively dual-plane designs with a wide range of variations offered over the years of production. Factory intakes were made for single 4- and dual 4-barrel configurations, as well as three 2-barrel versions. Most OE performance intakes were specifically designed for each port configuration. As development continued, Ford often made changes, both incremental and major, to the intake manifolds during production. As a result, you often see otherwise similar manifolds with noticeably different plenum shapes, sizes, carb mounting pads, etc.

The majority of factory high-performance intakes were made from aluminum, although some were cast iron, such as the 428 Cobra Jet intakes. There are a lot of used aluminum intakes available in the marketplace because they almost never wear out and are rarely damaged when other engine parts fail. These days a factory manifold's desirability, and hence price, is more dependent on the part number and date code than on any real or perceived performance advantage.

Low Risers

By far, the most common aluminum factory manifolds found in swap meets and auctions are the various low-riser pieces. Low-riser intakes are often reasonably low in profile and provide good hood clearance. They were available in a wide array of carb configurations and designs over the years, ranging from a single 4-, dual 4-, and three 2-barrel layouts. One thing to look out for (or look for, if you need one) is the intakes specifically used in Thunderbirds. These have a flat carburetor-mounting surface as compared to the normal angled pad(s) found on other FE parts.

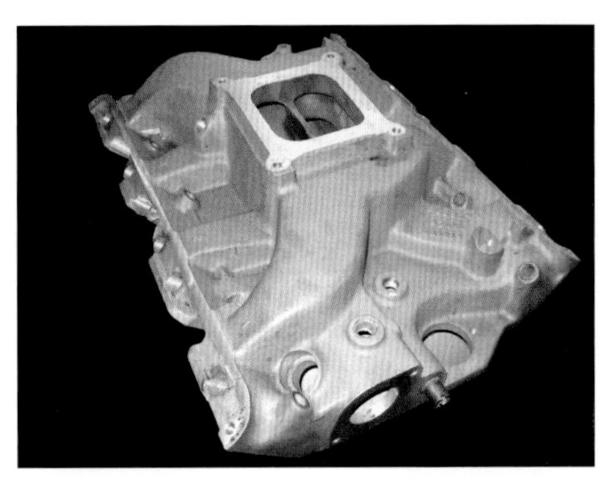

A factory sidewinder 427 intake is shown from the front corner. Early factory intakes have oil-filler provisions, while later ones do not. Only Blue Thunder offers an aftermarket dual-plane with the oil-filler opening already machined.

A NASCAR-era Ford experimental intake manifold shows you the engineering approach for increasing power in the 1960s. You won't see one of these at a local garage sale.

Medium Risers

Original aluminum medium-riser intakes are a bit less common but still readily found. Some of the aftermarket reproductions are so close to originals in appearance that it's hard to tell the difference. Certain factory 427 medium-riser intakes have a strong following among FE enthusiasts. In particular, the single 4-barrel "sidewinder," an intake where the carburetor was mounted visibly off-center, is very popular. The "Police Interceptor" intake is another aluminum single 4-barrel part that's worthy of note. Both of these remain popular upgrades for enthusiasts wishing to pick up power, reduce weight, and maintain a factory appearance.

This 2x4 carburetor setup sits on a Tunnel Wedge intake, and this classic good-looking factory-style combination still runs very strong today. I've had them at nearly 800 hp in marginally streetable combinations.

The "Tunnel Wedge" manifold is a unique item. Originally made available in the late 1960s only as an over-the-counter part from Ford dealerships, this intake essentially blended the tunnel port single plane concept with the common medium-riser FE port layout. Still available as an aftermarket reproduction from Dove, the Tunnel Wedge is a surprisingly current and solid performance piece, considering that the basic design is now over 40 years old.

Aftermarket Intakes

Most aftermarket performance manufacturers have made a multitude of intake designs for the FE engine family over the years. They include pretty much every conceivable variation on carburetor count, style, single-plane, dual-plane, tunnel ram, cross ram, high rise, low profile, 4150, 4500, and more.

Before going any further, I need to clarify the term "high rise." In the performance market, a manifold with a carburetor-mounting pad that is physically higher than the stock unit is often considered to be a high-rise intake. But do not confuse a high-rise manifold with a high-rise FE cylinder-head definition. You can indeed purchase a so-called "high-rise" intake that fits on low-riser heads, but a "high-riser" intake does not fit on those heads at all.

Performance aftermarket intakes from the 1960s emulated Ford's offerings with dedicated high-riser, low-riser, and medium-riser designs. Most current aftermarket intakes would be best defined as having a medium-riser port design with the raised floor. All of them incorporate a large enough gasket flange to fit on and seal to the more common OEM low-riser-style heads, as well as the medium-riser variations.

Aftermarket Dual-Plane Intakes

Dual-plane intake manifolds, with the port runners split into upper and lower groups and having a divided plenum, are the most popular choice for street-oriented builds. The dual-plane design has proven to deliver quick throttle response, excellent acceleration, and good driving characteristics. Virtually every OEM carbureted intake that was ever installed onto a production vehicle was a dual-plane design; that is no accident.

Used, older aftermarket dual-plane intakes are readily available at fairly low cost, usually far lower than the price of a comparable factory performance part. If you are working on a budget or are seeking a 1960s/1970s appearance for your engine, one of these parts is a viable selection. Performance potential is generally decent—certainly equal to or better than most factory offerings. The patina and cosmetics "fit" the look of many FE-powered vehicles. Although the dual-plane intakes are acceptable for a street build, these older aftermarket intakes are not the best choice for a race build. Some of these, notably the older Edelbrock

A Blue Thunder Weber intake is shown installed. This is only available for medium-riser heads, and is cast with a fairly small port. Many Shelby Cobra owners have used Weber intakes and carbs, and they are very attractive when installed. They are available in either a straight-mounting or a 15-degree inward-angled mounting for scoop clearance on a Cobra.

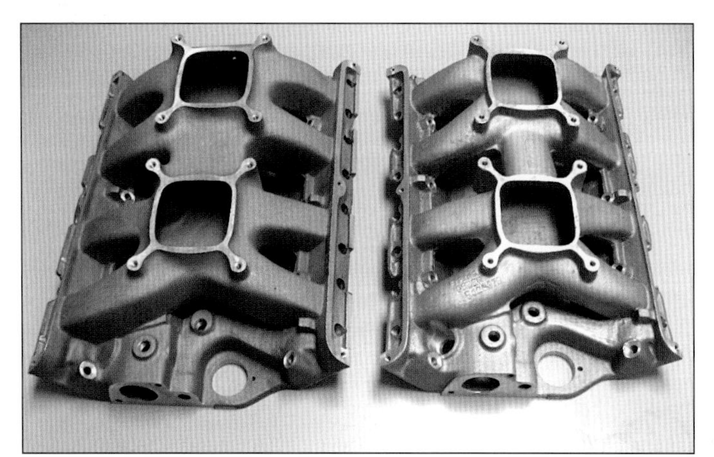

Dove high-riser versus medium-riser Tunnel Wedge intakes. What's interesting is that these Tunnel Wedge intakes provide exceptional performance while maintaining respectable street manners.

This is the rear view of the Dove high-riser and medium-riser Tunnel Wedge intakes. While similar in design, the difference in plenum size and height can be seen when the intakes are set alongside each other.

F-427, have a cult following of sorts and command a price premium at the swap meets.

Dual-plane intakes have undergone continuous development over the past 30 years, and the FE has been a beneficiary of those efforts. The current Edelbrock Performer RPM is among the best intakes available for any street-oriented project, as well as for many race engines. The Performer RPM performed extremely well in Jay Brown's comparative testing and has been proven to be capable of supporting more than 600 hp in my own dyno development work. For a project builder who wants an intake that is cost effective, readily available, and has power delivery potential, you cannot go wrong with a Performer RPM.

Another intake for Weber carbs; this one is unusual because of the uncommon sideways orientation.

In comparison, the basic Edelbrock Performer is something of a disappointment. It can best be considered as an option for somebody who simply needs a 4-barrel aluminum manifold. The cost is modest, availability is obviously good, but the casting design, port shapes, and machining leave quite a bit to be desired. Its sister, the Performer RPM, is a better part in every possible way.

The Blue Thunder dual-plane intake provides excellent flow characteristics and has a number of highly desirable features. Although it costs more than the Edelbrock RPM, it is a great choice for a 428 Cobra Jet engine, and it provides a reasonably stock appearance for any FE engine. It looks very much like an original Ford part with no visible brand logo. It is the only current aftermarket dual plane to include all of the oil fill and vacuum port provisions found in factory Ford intakes. The manifold-mounted oil-filler tube is necessary for use on 1965 and earlier engines with the correct factory valve covers. The normal 4150 flange Blue Thunder single-4 essentially matches the Edelbrock RPM for power potential, but this also offers the option of a Holley 4500 Dominator mounting-flange version that is suitable for big-cubic-inch stroker engines. I've run the Dominator version on engines approaching 700 hp at more than 6,500 rpm and had outstanding results. One thing to be aware of is that a spacer is required to use the universal Holley carburetors or the linkage hits the manifold.

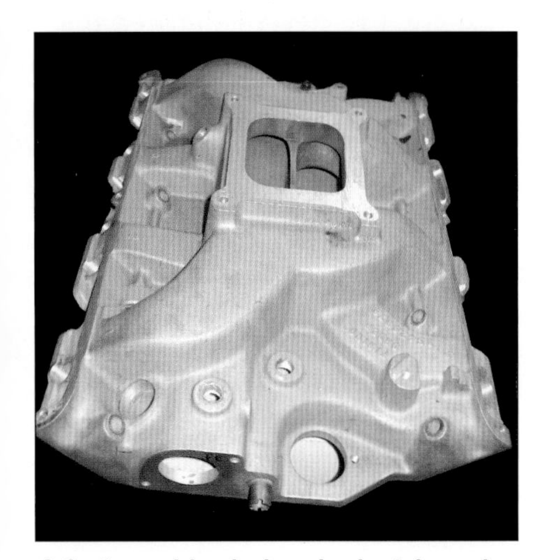

A factory sidewinder single 4-barrel intake manifold is offset-mounted but provides good flow for a factory piece. However, the newer performance intakes using better flow data, improved manufacturing techniques, and better materials yield better performance.

A Dove tunnel-ram intake is perhaps the best intake for maximum performance on naturally aspirated engines. In fact, a tunnel ram can provide exceptional performance across a wide RPM range. But unfortunately, these intakes are nearly impossible to find.

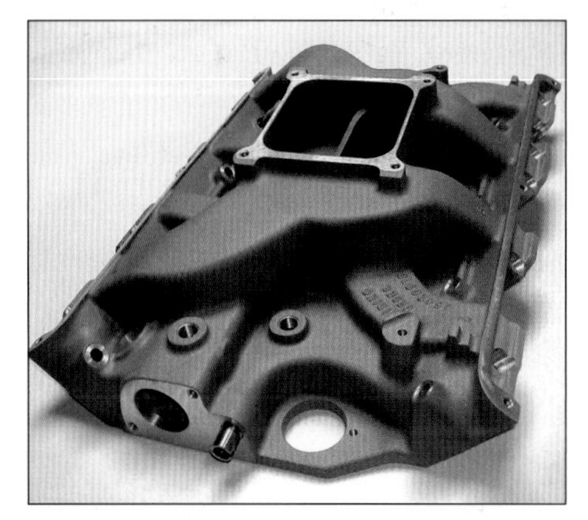

Blue Thunder FE intakes are offered in standard Holley 4150 or 4500 Dominator flange versions. The Blue Thunder dual-plane intake provides superb flow and it's the only aftermarket dual plane to include all of the factory oil fill and vacuum-port provisions. It is a great choice for any FE engine and has a close-to-stock appearance.

Aftermarket Single-Plane Intakes

The number of single-plane FE intakes is much smaller. A couple of them are essentially "leftovers" from the first gas crisis of the 1970s. Both Edelbrock and Holley released small-runner single-plane designs that were intended to be "the best of both worlds," with good street manners and power potential. The Holley Street Dominator and the Edelbrock Streetmaster are both worth chasing down as an alternative to new parts. Although the small-runner single-plane concept has fallen from favor, it turns out that these are both very good parts that deliver on their original promise. Testing has shown them to be comparable to the best of the newer dual-plane designs as manufactured, and race experience has proven them to be very competitive when properly modified. The Offenhauser Port-o-Sonic also has a strong following among FE racers as a base for extensive modification.

There are only a very few new single-plane intakes available for the FE. The Edelbrock Victor is by far the most common; the others are a group of specialty parts from Dove.

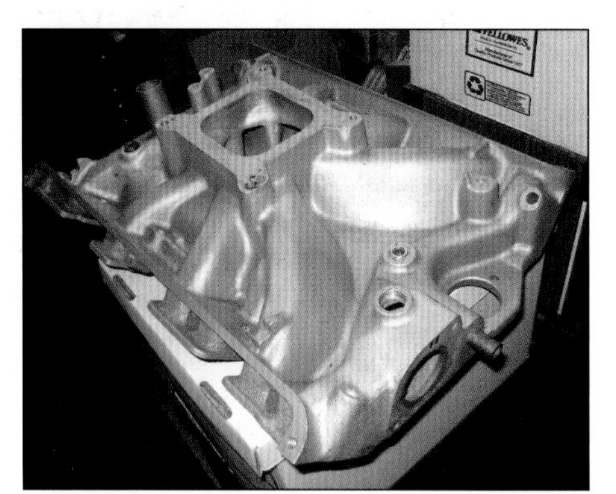

Edelbrock Streetmaster and Holley Street Dominator are single-plane intakes that share a similar external design. Independent testing has shown performance comparable to the best of the newer dual-plane intakes, and in racing these are very competitive when correctly modified.

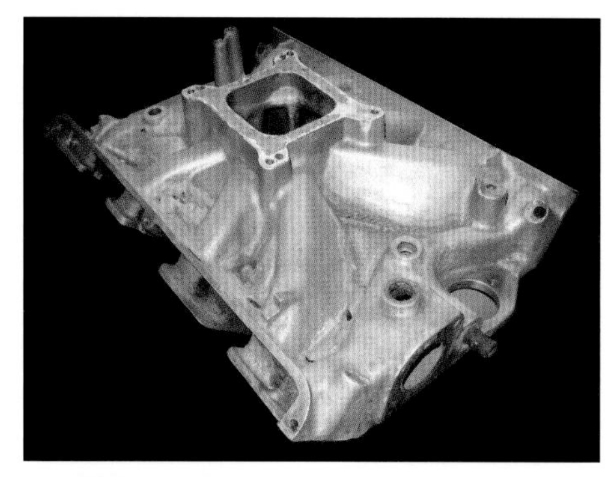

A 1970s vintage Edelbrock Streetmaster intake has the classic "spider" single-plane appearance. Contrary to popular opinion, these do provide very good performance on the street.

The Victor is available with either the common 4150 carburetor-mounting flange or a 4500 Dominator flange. Other than the carb base the intakes appear to be identical. In addition, the Victor can be purchased with EFI injector bungs already cast into place and machined for those wishing to convert their FE to port-style fuel injection.

While other Edelbrock intakes are ready to install as delivered, the Victor should be considered a base for modification. It can be simply unwrapped and bolted on (assuming everything fits (this is covered later), but the ports "as cast" are very small and would result in seriously restricted performance. The real design intent for this manifold was to deliver enough casting material for the dedicated builder to be able to port and modify as needed to complement the chosen cylinder heads. The Victor is, thus, a good candidate for a dedicated race engine. Just remember to include the port work in your budget.

The Dove intakes are also available in 4150 and Dominator flange designs. Being both taller than the Victor and having a larger plenum, they would intuitively be better on a large-displacement or high-RPM combination out of the box. Unfortunately, I've never compared them that way because I don't run the Victor without port work. The Dove parts were originally intended for racers and racers approach things from a different perspective, with a focus on light weight. Therefore, it has thin casting designs that don't allow for as much modification creativity without breaking through the casting surface. For many years, the Dove parts were the only single-plane FE intake available, and some folks have gone very, very fast with them. So they are certainly worth considering.

The Dove single 4-barrel is available with either 4150 or 4500 carb-mounting flange. Dove is a specialty racing parts manufacturer offering both high-riser and medium-riser port versions. Built for racing, these intakes have thin-wall castings to save weight, but the thin-wall design doesn't allow room for much modification.

If you are building a high-riser-headed engine and want a single 4-barrel single-plane intake, the only choice is Dove. The Dove high-riser single 4-barrel intake has huge runners, a cavernous plenum area, and is not likely to be a good fit for a low-RPM or small-displacement application. Strange as it may seem, you would be better off with the dual-quad setup, which has a smaller plenum and runner size for street use. The Dove units offer a couple different port designs: the traditional Ford port, and the "P.I.E.-"style port, which is named after perennial Super Stock record holder Ray Paquet. Ray continued the development of the high-riser Ford FE engine long after most others had moved on. Compared to the factory high riser, the P.I.E. port is somewhat squared off, has a dramatically filled floor, and is in some manner the inspiration for virtually every really high-powered FE drag engine built today.

Dual Quads

Multiple carburetors are virtually a signature feature on FE engines. That big-finned, oval, aluminum air-cleaner lid has become a universal icon and can even be found on non-Ford custom cars today. A well-tuned dual 4-barrel installation matches or exceeds the performance and drivability of a single 4. And nothing looks better under the hood of a classic Mustang or Galaxie.

A Blue Thunder 427 MR 2x4 intake is a faithful reproduction of the factory piece.

With the notable exception of the Tunnel Wedge, all factory dual-quad intakes for traditional FE engines are dual-plane designs. Performance differences doubtless exist between the various iterations offered up by Ford over the years, but power potential is rarely the selection criteria for these purchases anymore. The collector value of the factory-equipped cars has focused attention on correct part numbers and date codes, which is outside the focus of this book.

For the casual enthusiast, the main thing is availability and fit. Essentially you can make a low-riser or medium-riser intake fit on either head configuration. If you already own an intake or get a smoking deal, that's important. But, if you're out shopping, it is worth the effort to find one that fits the port without modification.

This leads us to the aftermarket. Currently available dual-quad intakes for the FE are generally based on near reproductions of factory parts, with only a couple exceptions.

Blue Thunder makes the most popular dual 4-barrel intake. The dual-plane design can best be described as being inspired by the factory medium-riser piece. The intake is cast as a medium-riser design but offers the option of a low-riser port match. A very nice part, the Blue Thunder 2x4 intake has enough casting material to allow port matching to highly modified heads without welding. It has the oil-fill tube provision in the front and the rear basket opening required for earlier cars, as well as the needed vacuum and water openings for various accessories.

The design enhancements that improve performance are not externally visible, so this is a great choice for a "resto-mod" type of vehicle in which cosmetics play a big role. The casual observer would be hard pressed to know that is was not an original intake on your car.

The Tunnel Wedge is the other popular dual 4-barrel intake; Dove offers it as a reproduction part. This intake is very close to the original in appearance, but interestingly, it was never a stock part on a Ford production vehicle. Instead, it was a dealer-only performance product that Ford offered over the counter. Since it was released near the end of the FE performance program, it was likely the most advanced multi-carb intake Ford ever made for the FE. The similarity of the Tunnel Wedge to modern performance intakes is startling, with the open-air gap below the runners. They were certainly well ahead of their time, lend themselves nicely to modification, and perform extremely well today.

Dove offers several versions of the Tunnel Wedge. The traditional medium-riser configuration is essentially a reproduction of the original Ford part, said to be made from the original tooling. Dove also supplies a high-riser head iteration with either the factory port or the "Paquet" port. I've used the latter design in engines reaching nearly 800 hp with minor modifications—a testament to the original design.

Edelbrock recently released a dual-quad Air Gap intake for the FE. Since the RPM single 4-barrel delivers excellent power, the dual quad should offer similar performance, but to a higher level. Unfortunately, Edelbrock chose to mount the carbs close together, limiting the intake to the AFB-style Edelbrock carbs. It's likely that FE racers will make modifications to mount larger carbs to these, and time will tell just how good the intake really is.

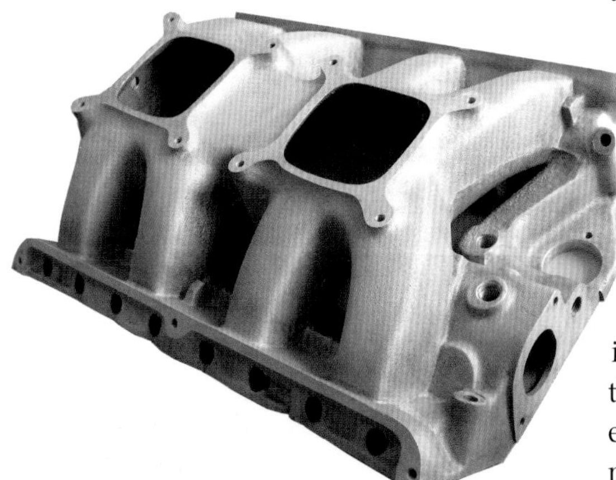

The Dove high-riser Tunnel Wedge is similar to the medium-riser factory part, but it's not the same. It's essentially a hybrid intake of the tunnel port single-plane concept with the common medium-riser FE port intake. This is an intake that Ford never offered.

A polished Tunnel Wedge makes quite a visual statement, and this provides exceptional performance for race and street applications as well. The intake certainly makes an impression when fitted to a Fairlane, Mustang, Torino, Galaxie, Thunderbird, or a street machine.

Other Choices: 3x2, Webers and Tunnel Rams

There are options out there for the FE builder who wants something different. Some of these are fairly available. Some are swap-meet or online-auction fodder, while others verge on the exotic.

The three 2-barrel combination is a common swap meet find, and these are typically mounted to factory intakes, which were installed by Ford for several years in the early 1960s on 390- and 406-equipped performance cars. They perform well on a high-performance street car and they have strong visual presence. But they are actually tougher to source than the dual-quad setup and nobody reproduces the intakes. Look out for T-Bird intakes, they are flat and won't look right on a Galaxie.

The FE Weber intakes are still available from Blue Thunder and are visually stunning when assembled. A favorite of the Cobra community, they are offered in either a straight-mounting or a 15-degree inward-angled mounting for scoop clearance. Inglese sells a package with the Blue Thunder intake and carbs ready to install.

TWM (among others) offers an EFI conversion package that delivers the Weber multi-stack appearance with the advantage of electronic fuel management. This package only includes the throttle bodies, intake, fuel rails, and linkage, so you need to source the management electronics and injectors to build a functioning system—very expensive. However, this is a viable option if you are seeking to make an instant visual impression while retaining good driving characteristics. I've run one of these on the dyno and it ran very well.

With the aftermarket FE 6x2 Intake, getting a smooth operating throttle linkage and routing fuel lines are a real challenge. It's a lot of parts: six lines and six pieces of linkage.

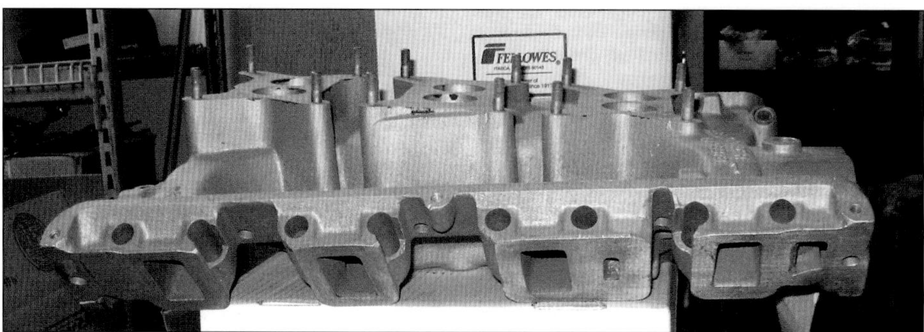

Galaxie 3x2 intake shows the carburetor-mounting angle, while the Thunderbird intakes are flat. This is because the FE engines are mounted on a downward sloping angle when installed in most Ford passenger cars. Thunderbirds have the engine mounted low and flat to minimize the intrusion of the transmission tunnel into the interior.

Dove made tunnel rams for a very short time and they are quite rare. They perform surprisingly well, look very cool, but are hard to find.

Edelbrock and Mickey Thompson made cross-ram intakes for the FE. These featured two 4-barrel carbs, which were offset—mounted essentially side by side. Unfortunately, they are not a particularly high-horsepower package, but are visually impressive, particularly on a period-style muscle car. If you find a Mickey Thompson intake, make sure you get the required distributor spacer as part of the purchase, so it can be properly installed.

Intakes with a variety of small 2-barrels were also available. Six 2s look pretty darn cool on an old-school street rod. You probably won't be fast, but you will get to work out your fabrication skills on fuel line plumbing and linkage.

Hilborn and Algon made stack-style injectors for the FE in the 1960s. These are not at all common today. And when you do find them they are expensive, often incomplete, and still as challenging as ever to set up. The option to convert them to EFI is there and would be impressive to see, but bring your checkbook and fabrication skills.

Blue Thunder offers supercharger intakes for mounting the GMC Roots-style blowers. They required a separate spacer between the blower and the intake for distributor clearance. Not many FEs are seen with the blower through the hood, but these supercharger setups are becoming more popular because Pro Street is making a bit of a comeback. But to successfully install such a setup is another fabrication exercise for the drive components. Hampton Blowers still advertises the parts as being available and may be a good resource for those desiring that package.

Here's a very expensive and exclusive fuel-injection system. TWM fuel injection system, based on Blue Thunder's Weber intake, provides a multi-stack carb appearance. The system includes throttle bodies, intake, fuel rails, and linkage, but you need to source the electronic management system from another supplier.

If you're taking the forced-induction route and installing a Roots-type supercharger, the Blue Thunder FE 6-71–style blower intake is an excellent option. A spacer must be installed between the intake and the blower to provide adequate distributor clearance.

Pressure Checking

I pressure check any intake where I've made significant modifications because past experience has shown that issues often crop up.

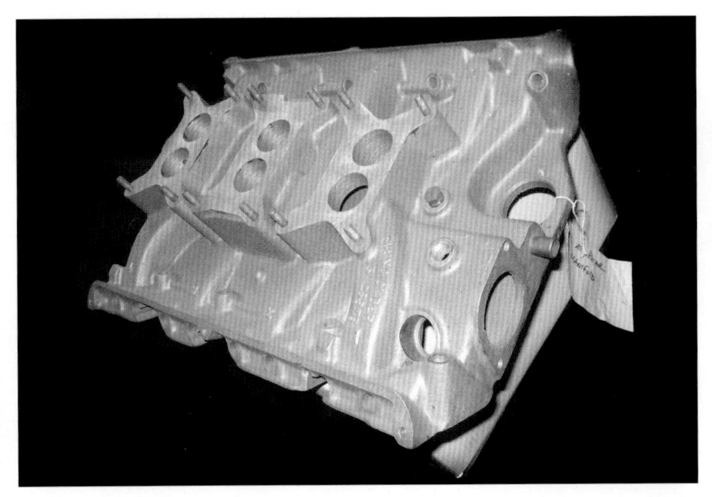

This is an alternate view of the 3x2 factory intake manifold for a Galaxie.

The FE 6x2 intake (shown from the front) is a popular choice for restoration or for an original-equipment look. It's a cool nostalgia piece and not a race part.

With the FE oil-filler tube, the end with the groove goes up and the end with the chamfer goes into the intake.

The pressure-testing rig is simple and easy to make. I use a block of 1/2-inch aluminum stock to make a pair of block-off plates—one for each side of the manifold. These only need to cover the water passage, and are mounted to the intake using the front two fasteners on each side, along with a gasket. The thermostat opening is blocked off with a similar fabricated plate and gasket. I plug all but one of the pipe-threaded water openings, usually using the bypass hole in the front as the place to apply the pressure. Pressure comes from a normal pump-style radiator pressure tester. A dozen pumps and some soapy water are all you need to find any leaks.

This simple rig has found flaws that could have caused a lot of trouble, including cracks through pushrod holes, where the oil-filler tube presses in, or porosities in castings that were otherwise invisible.

Bolting it On–Making it Fit

Perhaps the most challenging part of FE engine assembly involves the intake manifold. The FE intake, due to its unique design, interfaces with numerous other engine parts and can be a real challenge to properly install. If everything is correct, it is easy to install, but if components and machined surfaces do not line up with the intake, it can become a true nightmare if you're not paying attention. Some guys just get lucky, pull the intake out of the package, bolt it right down, and it fits perfectly. I'm not one of those guys, so I'll detail the arduous process I follow for intake installation in case you're like me.

If you combine the number of machined surfaces, the size of the intake, and the age and unknowable history of the various parts, you have a recipe for a very challenging assembly project. I've seen every sort of fit problem, such as pushrods that won't go through the holes, gasket flanges that are on multiple angles, valve-cover rails that don't line up, distributors that won't go in, bolts that won't fit, broken castings, and leaks. Yes, lots of leaks.

Purists do not like the idea of milling intake manifolds. Their position, and it's a valid one, is that you are making the manifold unusable in future engine builds with other blocks or heads. Personally, I will do anything needed to make one fit because my concern is the engine

Crankcase breather basket installs in the rear of an early-style intake. This is used in concert with a breather element or road draft tube.

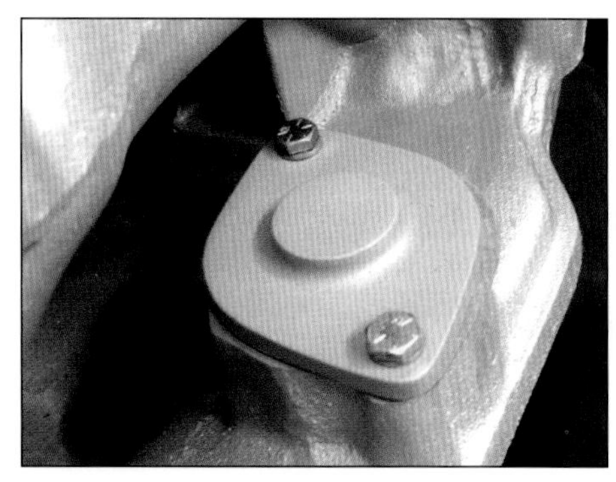

A solid cover is used on intakes when a breather is not required.

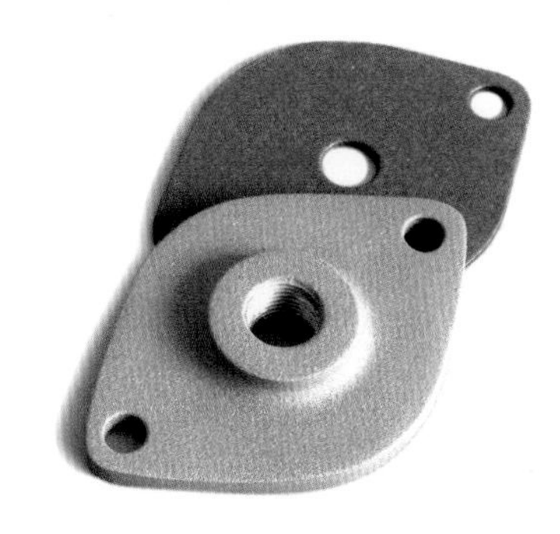

A breather cover has an integral tube, and provides an alternate hookup to a PCV line or remote crankcase vent.

The breather separator keeps oil mist from leaving the engine.

Intake bolt-hole alignment is important to prevent manifold damage. Improper alignment results in intake cracking at the boss, as well as stripped threads.

A simple pressure-checking rig can be created using a radiator-pressure tester and some aluminum stock. Pressure testing ensures that you do not have any coolant leaks to contaminate the oil.

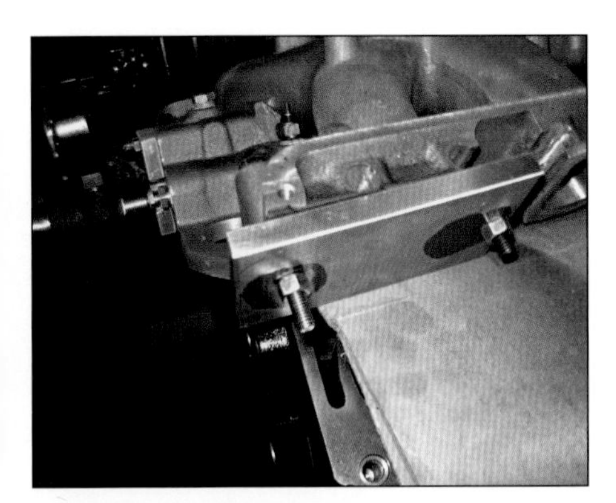

Pressure-checking plates are fabricated from 1/2-inch-thick aluminum stock and need only cover the water openings in the intake.

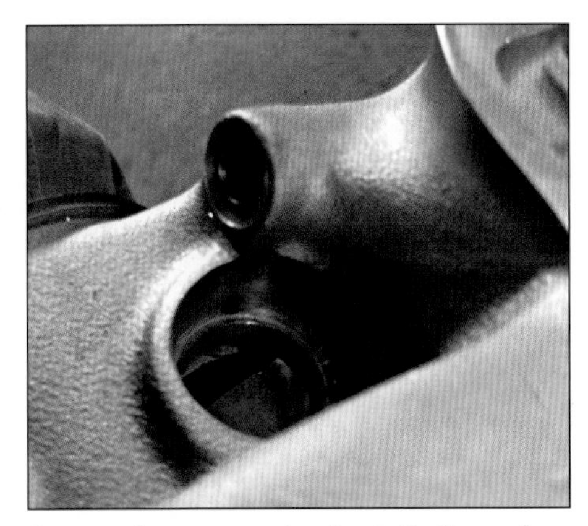

An overly aggressive installation of the filler tube led to a water leak that was found in the oil-filler tube opening. So, lesson be learned, don't use too big of a hammer!

On this intake, we found a leak at a casting parting line in the water jacket when pressure testing the coolant system.

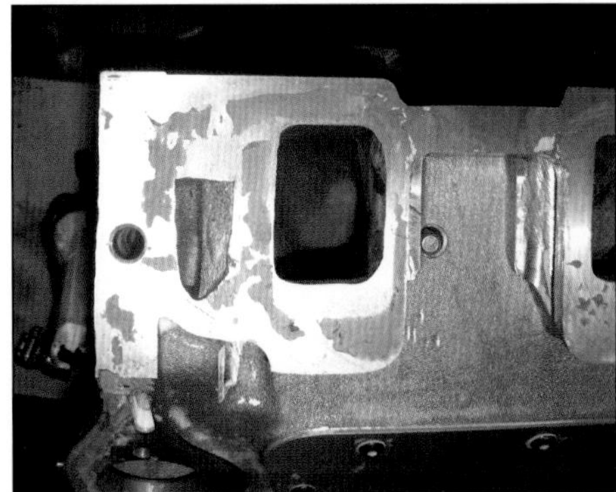

On some Dove intakes, this water coolant passage is exposed to the outside of the gasket opening and will cause a water leak if not addressed. If the opening goes behind the gasket you need to fill it with a bit of weld or epoxy to prevent a leak.

Correctly installing the distributor shaft through the intake manifold can be a little tricky. When the gap is even all the way around the shaft, the distributor is correctly centered in the manifold.

The proper distributor fit in intake is viewed from another angle. If the distributor shaft is off-center, you may not be able to turn the distributor for timing adjustment, or you may get an oil leak.

that I am building now and all the components need to be compatible with one another. I am not very concerned about an undetermined build in an unknown future. Here is how we do the job:

Milling Intake Manifolds

Milling the intake should provide a reliable seal for three surfaces: gasket flange, valve cover, and front and rear ends.

1 First check your bolts. The various FE intake manifolds have a variety of required fastener lengths. If the bolt does not have enough thread engagement or bottoms out in the cylinder head's holes, nothing is going to work correctly.

2 With all surfaces clean, lay a set of intake gaskets in place on the heads. I use Blue Thunder or Mr. Gasket parts for this step because they have no embossed sealing beads. Do not use any end seals at this time.

3 Lay the intake manifold in position, being careful not to disturb the gaskets. Use the flat of your hand or a plastic mallet to move the intake around until you can slide the distributor into position. The distributor serves as the front-to-back and side-to-side locating dowel; you are not installing it "for keeps" yet, but you want to center the hole in the intake around the distributor's seal as closely as possible.

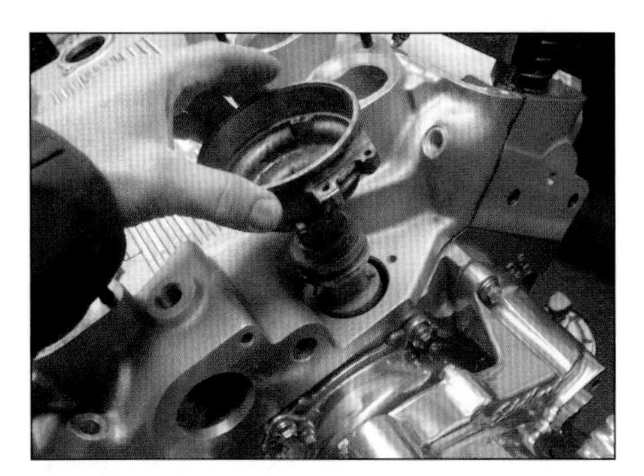

Installing the distributor in its proper location is a key to getting proper intake fit.

4 Next, look at the intake from the front and rear and use a stack of feeler gauges to level the end-seal surfaces from side to side. I call them the "China walls," and typically, I see a gap of around .120 to .170 inch. There is no critical number (you need to have a decent-sized gap) but anything under .030 will probably bottom out before the intake gaskets seal up, and this requires you to repeat the process later.

Here we are measuring the gap at the front "China wall." You want this to be at least .060 inch and level. Too tight and it can prevent the intake from sealing.

5 The intake is now properly positioned on the engine. Next take a flashlight, a dental mirror, and inspect for port alignment and bolt-hole alignment. You should be able to see the bolt-hole threads in the heads centered in the manifold openings. The intake and head port sides and roof should all line up. This almost never happens, unless you've already gone through this part of the process when porting the intake or if all the parts are new and to spec.

6 If both the bolt holes and the ports in the intake are higher than those in the heads, it is a fair bet that the intake gasket surfaces need to be milled. The FE head gasket and intake gasket surfaces are 90 degrees relative to each other, so any machining operation to the heads or block deck will have had a corresponding impact on the intake fit. Write your findings on the intake with a Sharpie-type marker—you'll need the data later.

7 Next take feeler gauges to the front and rear corners of the intake sealing flange of the heads. Probing between the intake gasket and the manifold, try to slip in a .010 feeler gauge in the top and bottom of each corner—front and rear, left and right. You are checking for matching angles between the intake and the head. If you cannot fit anything larger than .010-inch at any position, it's good. Now, it's time for the Sharpie again. On the respective corner of the intake, write the size and location of each variance. An example might be .020 inch at the top and .005 at the bottom of the left front.

We do our intake flange checking with a feeler gauge, and inspect front and rear, top and bottom. Warped and poorly machined FE intakes are extremely common, and cause a lot of post-build problems with poor running and oil consumption.

Checking the intake manifold fit at both the bottom and the top edge reveals any angle variances compared to the head. Problems here are far more common that you'd expect.

8 Take a quick look at the valve-cover rail. If the part on the intake is higher than the part on the head, use a dial caliper or a feeler gauge to find out by how much. If it's more than the amount to be removed from the gasket face, your machinist needs to trim that surface as well. For example: if you take .020 inch from the flange and the rail is .040 too high, it's still going to need an extra .020 inch taken off to get level to the head's corresponding rail surface.

To check valve cover rail alignment, you need to get the gasket and rail as level as possible. The gasket can only accommodate a small amount of variance.

9 If your first group of measurements noted that the intake had to be milled by .020 inch to line up the bolt holes, and your second group indicated that it was looser at the top by .015 inch due to an angle variance, you want to cut that gasket face at an angle—removing .015 inch at the top and .020 inch at the bottom. This fixes both problems in one cut. If you are making big angle changes, check the spot faces on the mounting fastener holes too. An angle contact breaks the mounting-hole boss, necessitating a common repair seen on older intakes.

Take off the manifold and sit down with the person who'll be doing the work. Show all your scribblings, and explain how you arrived at them. If he or she won't stop and listen, and understand, find another machinist because you cannot put the metal back on once it has been removed. I tend to draw lots of pictures and angles, along with putting all my measurements on the manifold itself. It makes it easier to avoid misunderstandings.

10 After machining, clean out all the chips and reassemble the intake onto the engine. The bolts should line up now, and can be lightly screwed into place. Don't use any adhesive or sealant because you are not done yet.

11 Pushrod fitment is the next step. Dry assemble the valvetrain with the lifters, rocker assemblies, and pushrods. Do a rough set on valve lash. Turn the engine over and watch the pushrods carefully to see if any of them hit their respective holes in the intake. Most mild combinations are okay, but engines with larger cams cam have an interference problem. The dental mirror and flashlight come in handy again because contact can be at the lower end of the hole where it's tough to see. Sometimes it's helpful to slip a light below the intake through the distributor or rear breather openings. If nothing hits, you can proceed with the assembly. If it does hit, mark the offending spots, remove the manifold, and grind for adequate clearance. This is a case in which it's okay to get close, but not to touch. If you grind through into an intake port, you need to weld it up or install a sleeve into the pushrod hole. Brass tubing stock from the hobby supply shop works nicely when retained with a bit of epoxy.

Note the intake pushrod hole clearance. You want to rotate the engine through several cycles during test assembly to verify smooth operation of every pushrod at all valve-opening positions. Hard contact when the engine is running causes the pushrod to pop off the rocker arm. And you definitely want to avoid that.

This is a detail photo of pushrod tubes in intake. In order to install the pushrod tubes, a light push and the use of some epoxy are all that's required.

Port Matching and Beyond

With the wide variety of intakes and heads available for the FE engine family, it's almost a given that a certain degree of modification is required during assembly of the top end. I do the port alignment and matching work at the same time as the fitting process.

1 My port-match technique is pretty straightforward and goes quickly once you've tried it a couple times. I port match to the heads but not the gasket. Gaskets get trimmed as needed. First, make the intake fit (as discussed above). You can't port match the intake if it then has to be moved in order to fit.

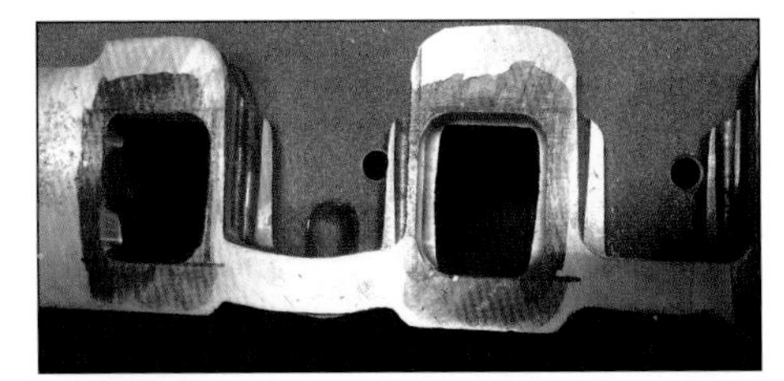

In the first step of port matching, use a Bridgeport to rough-in the new enlarged openings. You don't want to go beyond the edge of the guide marks, so be careful. Remember: If you remove too much material, you cannot add it back on without a welder. You just need to get close to the edge, so you can hand finish them to fit.

2 Next paint the intake manifold and cylinder head gasket faces with machinist's dye and let them dry. Be certain to put 1/2 inch of the dye around the front, rear, and top of the head—that's where your transfer markings will be.

3 Now use a straight edge and scribe to mark a line delineating the top and bottom of the ports along the head's intake gasket surface. Use a small square to draw vertical lines along the sides of each port extending to the top of the gasket flange. Use your scribe or a Sharpie to extend these vertical and horizontal lines onto the dyed area around the top, front, and rear of the head.

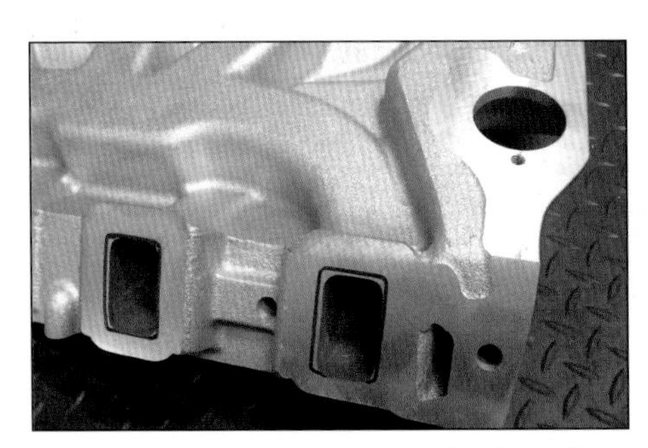

The intake is properly marked so the intake and its gasket correctly align. I used a Sharpie to illustrate how to align this intake. Typically, I scribe a layout line after coating the surface with machinist's dye.

Some intakes do not accommodate large ports without welding and a bit of rework, but this fabrication work is only seen from the inside and not noticeable once the engine is assembled.

4 Next lay some gaskets in place and mount the intake as normal using the distributor and a couple bolts. There's no need to tighten them at this point. Using the marks you just made on the heads, extend them onto the matching front, rear, and upper surfaces on the manifold. Remove the intake.

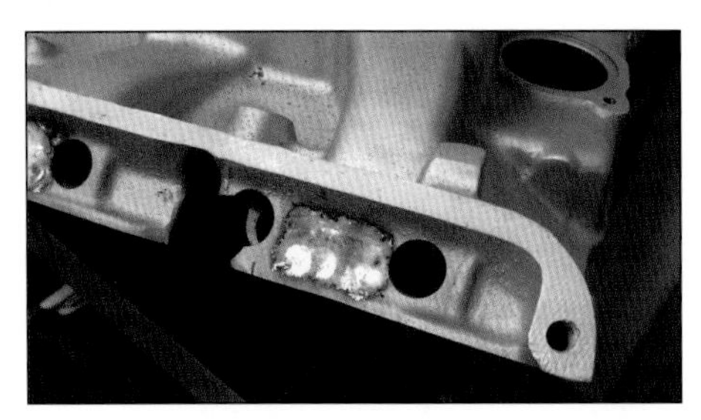

When porting the intake manifold to accommodate race heads, the factory and Dove intakes often have a small indent in the casting above each port that may need to be filled with weld to allow a good match.

5 The marks can now be "wrapped" around to the intake's gasket flange and drawn in the dye using the straight edge. At this point you've located all the head ports vertically and horizontally, which is all that's necessary to measure and duplicate the corner radius.

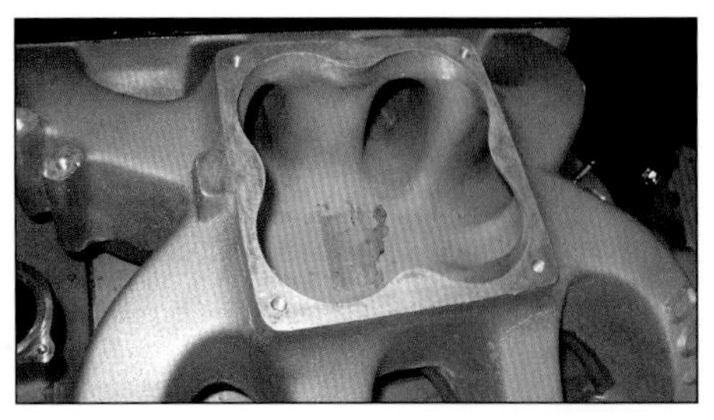

The extensive plenum modifications on a Dove single-four high-riser intake was done to dramatically lower the operating RPM range of the intake for a specific competition environment.

Every porter finds a preferred corner radius on a particular head and duplicates it on the other ports of the head. I use anything handy to measure and duplicate their chosen corner radius. A selection of deep-well sockets will usually work nicely, and those are always nearby. Scribe the matching radius into the dye at each port outline, and your map is complete.

When shaping the ports of the intake, I use a Bridgeport mill to get the first 1/4 inch of the opening shaped, but it can be done very nicely by hand with a die grinder. Use a carbide burr that is made for aluminum, it has wider spacing between the cutting teeth and won't clog with aluminum shavings. A protective layer of masking tape around the port opening on the gasket flange comes in handy if you're nervous. Go slowly, use a steady hand, and remember that the idea is to get a smooth transition from the intake into the head. No big bumps, no lumpy shapes, and don't try to "bellmouth" the runner. Be really cautious around the pushrod openings. A sanding roll on the die grinder can be used to straighten, finish, and polish the port surface.

There are cases in which the port opening extends beyond the casting surface. To resolve this problem, weld before grinding the opening.

Certain factory and Dove intakes have a small depression cast into the top of each runner just above the port. At Survival Motorsports, we sometimes fill that area with weld before working on the ports. Similarly, sometimes we cut through the pushrod tubes and use the welder or brass tubing as a fix.

We let our head porter do any radical modifications, but smoothing and blending the casting in the plenum and entry is worth doing. Don't go for sharp edges; rounded corners, gradual entries, and smooth transitions. That noted, only the engine tested in the field on a dyno can tell you what works best. I've seen some really horrible-looking work that ran great, and some really pretty stuff that was not up to expectations.

Weld-in EFI bungs are added when converting a carb intake to injection. Only the Edelbrock Victor is available with EFI bungs already cast into the intake.

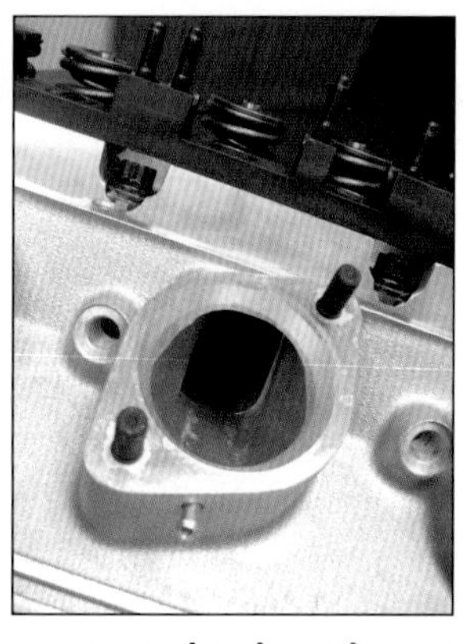

Port alignment is important if you are going to get the most out of your cylinder heads. Here you see an edge hanging out into the airflow path—an undesirable situation.

Final Installation

You can (finally) complete the assembly once the procedure has been checked, you know everything lines up and is square, and no components are hitting one another. My preferred gaskets are either Blue Thunder or the newest steel-lined Fel-Pro parts. Blue Thunder gaskets are easier to trim for custom port shapes but Fel-Pro parts are more readily available and have integral sealing beads. Avoid the traditional Fel-Pro Printoseal race gaskets in street-oriented FE builds. The reason is that the FE's unique intake surface is entirely surrounded by oil, and the non-reinforced race gaskets absorb the oil and fail in continuous street service.

1 I use Motorcraft TA-31 diesel application silicone as a sealer because it is highly oil resistant, dries to a near plastic like consistency, and has a gray color that looks good against an aluminum intake. Put a thin "finger wipe" layer on the front and rear end seal surfaces first, and then apply a thick bead. The thin layer assures good adhesion to the block and intake. Use a very, very thin layer that is like a coat of paint (not a bead) and this should be the same sealer on the water and port openings. With the FE's tendency to leak, a little insurance is a good thing.

2 Install the intake while the silicone is still wet. Now, you want it to stick, right? Move the intake around just like the test fitting, using the distributor as a locator.

3 Install all the fasteners loosely before tightening any of them. You may have to tap the intake around to get them all into place so you don't need a tight one holding it. If you are installing factory-style dual quads, don't forget to install the fuel log under the appropriate bolts.

4 Once all components are correctly positioned, you can tighten the fasteners. The factory manual has a torquing pattern, but you can torque down the fasteners from the inside out in a spiral. They do not need to be really tight, just snug, and it's nearly impossible to put a torque wrench on many of them, in any case. If the fit is correct you are often able see a really, really tiny bead of silicone showing around the outside of the port openings by looking down into the valve cover area.

5 Next pull the distributor back out. Usually you need to clear a bit of excess silicone from around the opening. Let the silicone dry (overnight is best) and use a sharp Xacto or hobby knife to trim the excess from the front and rear edges. If the intake gasket protrudes above the valve cover rail, you should also trim that level.

Finally—you're done.

CHAPTER 11

IGNITION SYSTEMS

A factory FE distributor with a Duraspark electronic conversion and a large cap provides far better performance than the strictly stock setup. The benefit of using factory parts lies in the ability to easily find replacements if required.

Ford FE engines have several viable performance options for ignition. They are good ones that are equal to those available for any other engines. All FE-equipped vehicles (with rare exception) had a traditional points-style distributor and ignition system from 1958 until 1974. That rare exception was a short-lived transistorized system for race 427 engines. You won't find those outside of the collector's environment. In 1975 Ford went to electronic ignition using a magnetic trigger distributor and the Duraspark system.

Factory Points Distributors

The vast majority of Ford FE engines came from the factory with a single-point distributor. Although there were a variety of advance curve calibrations from the factory, these do not have much relevance to the high-performance builder. The same can be said for the various casting numbers, date codes, and designs.

The high-performance factory engines were available with a dual-point distributor and no vacuum advance, but these are rare items. And while they run well when freshened up, they are best suited for restoration work. In reality, breaker-point ignition cannot match the consistency and performance of the CDI ignition system. Therefore a current MSD, Mallory, or similar ignition system is recommended for a high-performance build.

Ford FE distributors actually mount into the engine block and simply go "through" the intake manifold. The block has a small recess just above the cam for indexing distributor housing. The opening in the intake manifold is sealed from oil spray with a cone-shaped rubber O-ring installed on the distributor body. The distributor hold down, while mounted to the intake, actually clamps the distributor housing against the recess in the block, but not the intake itself.

At the bottom of the block's distributor opening, below the cam gear, is a smaller-diameter guide hole for the end of the distributor shaft, which helps take up any side loads from driving the oil pump. This opening is larger on medium-duty

"FT" truck blocks, which necessitates the use of an adapter bushing to run a passenger-car distributor. The bushing was available from Ford, but it has since been discontinued. Therefore, a bushing would have to be fabricated if required.

All Ford factory distributors for the FE used the traditional small-diameter distributor cap and push-in-style plug wires. A larger-diameter cap with more modern plug-on wiring can be readily adapted using parts from later model small-block engines. Unfortunately, the large-diameter cap does not clear some carb and air cleaner combinations, notably the popular 2x4 setups. Those barely have enough room for the standard-diameter cap to fit.

The mechanical-advance mechanism in original Ford FE distributors lies beneath the lower points/pickup mounting plate. A vertical tab that contacts a slot/relief limits total mechanical advance. A pair of springs, which work against pivoting weights, control the rate of advance. To change the mechanical—advance curve rate and amount, you must alter the springs and weights, and also modify the slots or tabs.

Ford vacuum-advance systems work by pulling the points upper mounting plate around a pivot on the fixed-position lower plate. Earlier vacuum-advance canisters have a bolt-in steel line rather than a hose connection. Depending on the year and vehicle, the later hose-connection canisters may have had either a single or a dual diaphragm. The dual-diaphragm units were emission oriented, and the two connections are rarely used in performance applications, so you plug the retard side. There were varied amounts of advance designed into the original-equipment canisters; some were adjustable using a hex key inserted into the vacuum hose opening. But 40 years of parts swapping, junkyards, parts store rebuilds, and general service have rendered that application data invalid unless you have a known original and untouched part.

Factory-Style Electronic Ignition

Rebuilding an original-points distributor with new bushings and such delivers a reliable ignition for a stock or near-stock-type build. But ignition system technology has long since

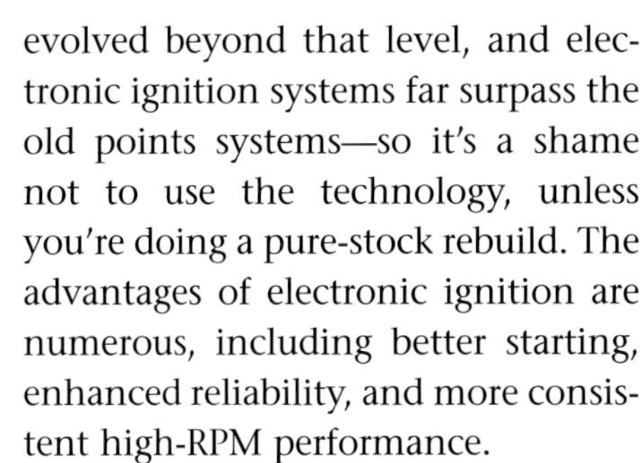

evolved beyond that level, and electronic ignition systems far surpass the old points systems—so it's a shame not to use the technology, unless you're doing a pure-stock rebuild. The advantages of electronic ignition are numerous, including better starting, enhanced reliability, and more consistent high-RPM performance.

The Ford truck distributors from 1975 and 1976 have the same electronic pickup coil and harness connection as other Ford engines of the period. These can be retrofitted to earlier engines and either used with factory-style ignition systems, or to trigger an MSD capacitive-ignition box. These are still available as rebuilt units from most local auto parts suppliers and it's the most cost-effective, solid-performing electronic ignition for your FE engine that uses original Ford parts.

Electronic Conversions

There are a couple of ways to go electronic without looking "new," which is important on a cosmetically period-correct hot rod. The easiest is to simply run the Duraspark distributor as noted earlier. Although obviously not original, it still looks like a Ford part, because it is one.

The next option is a Pertronix conversion kit. In the 1970s there

This is another early distributor with the Duraspark conversion. Although it looks stock from the outside, internally, it uses modern capacitor discharge ignition components to provide the strongest, most reliable spark.

The small spacer is used with shims to set the end play in the distributor.

were a number of similar kits on the market that let you turn a points distributor into an electronic distributor. Pertronix is the only large company that still services that market niche, and it has continued to refine the system over the years. The kit simply replaces the points, and does a nice job in street-oriented builds. You need to remember to retain the ground strap connecting the upper and lower points mounting plates; it is the only way that the Pertronix system gets a usable ground.

Another option is to combine the Duraspark internals with the original Ford points-type distributor housing. I have done some stealthy conversion work to retain the original external appearance while using modern electronic internals on dual-point or older original-equipment-type combinations. But there is one thing to keep in mind: If you're using an OE points-type distributor as a base to work from, modification of a rare or valuable OEM distributor will render it worthless to a collector. It's a good idea to check out that part number before you start.

Aftermarket Distributors

FE engine owners always had a variety of distributors to choose from in the performance aftermarket. Mallory, Accel, MSD, and a few smaller companies offered single-point, dual-point, electronic, and even magneto ignitions. While the points-type systems have faded away, MSD and Mallory still offer high-quality billet parts that are more than up to the task for any build.

Importers have filled the low-cost void with inexpensive distributors that mirror the appearance of high-end billet parts at a fraction of the cost. I have no experience with any of these and do not recommend using them. Like most automotive equipment, you get what you pay for and trying to cut corners with an ignition system can be very harmful to an engine.

While an old school aftermarket dual-point or mag might be the visual "ticket" for a vintage race-car look, you need to consider that service items, such as caps, rotors, and internal parts are going to be very hard to find. Cool has its price.

For milder applications, MSD sells a "ready-to-run" distributor with the ignition control module built in. All you need to do is connect the harness as instructed, which requires a ground and 12-volt "key on" signal. The unit includes an adjustable vacuum advance and is as simple as it gets for a street-rod ignition package.

An MSD billet distributor is the most popular choice for a hot street- or race-type application. The heavy-duty distributor body is machined from a solid bar of 6061 T-6 aluminum. This provides reliable performance, and is said to be able to handle up to 10,000 rpm—well beyond the working range of any FE. The mechanical advance assembly is positioned so it can be adjusted without disassembling the distributor.

Mallory sells a large-cap Unilite distributor with an optical trigger, rather than the more common magnetic trigger used by Ford and MSD. Although the Unilite is certainly a step up compared to an older factory points system, this package is less popular than the MSD, and hence has fewer options in terms of compatibility with other parts. While I have limited experience with them, others report good results, and it is likely to be a reliable alternative. One note: The large cap may preclude its use with the tight carb clearance of the dual-quad intakes.

For serious street and race engines, the first choice has become the MSD billet distributor. It's readily available, durable, and very easy to adjust and tune. The MSD billet uses a small-diameter GM-type cap, but it's a tight fit and just barely clears the front carb bowl on 2x4 setups. A Ford style electronic pickup coil is internally mounted to a non-moving plate. Only two wires extend from the distributor: a purple and green one for the magnetic pickup. An important thing to remember is that the purple MSD pickup wire *does not* connect to the purple wire in an original Ford

The MSD billet distributor advance mechanism is GM-style, and it places all the parts up top, so they are easy to modify. Nylon pads support the ground-advance weights for smooth operation.

distributor wiring harness. MSD clearly states this in the instructions, and some owners still connect the two purple wires. Getting it wrong will have a dramatic affect on ignition timing. The MSD pickup leads terminate into a small plastic "AMP"-style pin connector. I've seen folks use a small wire tie to ensure that the distributor and vehicle harness remain connected. I prefer to change the distributor to a self-locking and moisture-resistant WeatherPack-style connector.

The advance weights and springs in an MSD billet distributor are located just below the rotor, similar to the GM configuration. The total advance as well as the rate can be changed in a matter of minutes with the distributor still mounted in the engine. Locking the advance out completely is done off the car, by removing the distributor shaft and rotating it 180 degrees. Once reassembled, the advance is rendered non-functional.

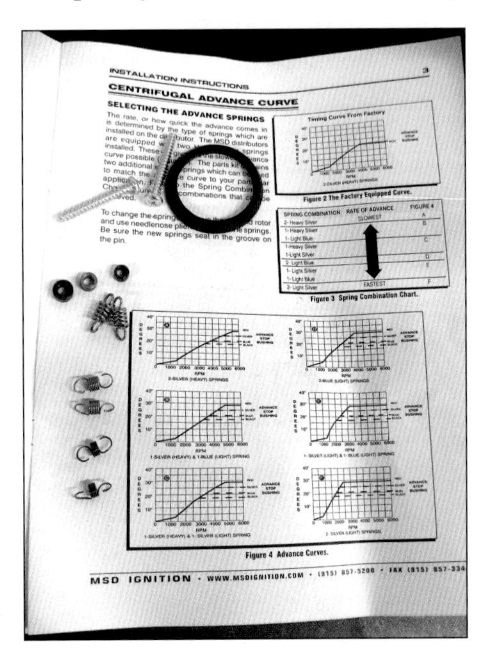

MSD distributor advance-curve parts come with the billet distributors. This includes a ground-advance cam that's tig-welded to the 9/16-inch hardened shaft. The shaft diameter mandates the use of a 460 Ford .531 ID distributor gear.

Although the billet unit has no vacuum advance, I won't miss the feature on race or extreme high-performance engines. With the MSD billet unit, as compared to the ready-to-run unit, you have the advantage of choosing the type of ignition control box, and I usually opt for a higher-powered MSD 6 or 7 series. These are capacitive discharge ignition boxes that dramatically step up the coil primary voltage—hundreds of volts compared to the old points systems' less than 12 volts that. The combination of a coil designed around this ability, better wires, more energy, and electronic ignition's greater accuracy delivers a faster and stronger spark than was ever possible in the good old days.

Alternately, you can use only the distributor features and bypass the pickup coil entirely to use a crank trigger. The crank trigger is the most accurate means of controlling the spark timing signal. MSD discontinued the FE crank trigger kit a couple years ago, but one can be easily fabricated. The trigger wheel for a small-block Chevy has the right diameter and can be mounted to the damper with only minor bolt-hole tweaking. A fabricated magnetic pickup mount can be attached to the two lower holes in the passenger side of the timing cover. The trigger signal can either be used directly by an ignition box to fire the coil, or it can be modified through use of a timing-control computer.

Plug Wires and Plugs

The demands put on spark-plug wiring for an FE is no different than those of any other engine. As spark energy has increased, so have the requirements for a really good plug wire. Most aftermarket suppliers have kept pace with this need by offering lower-resistance, helically wound wire with large sturdy boots and thick layers of insulation.

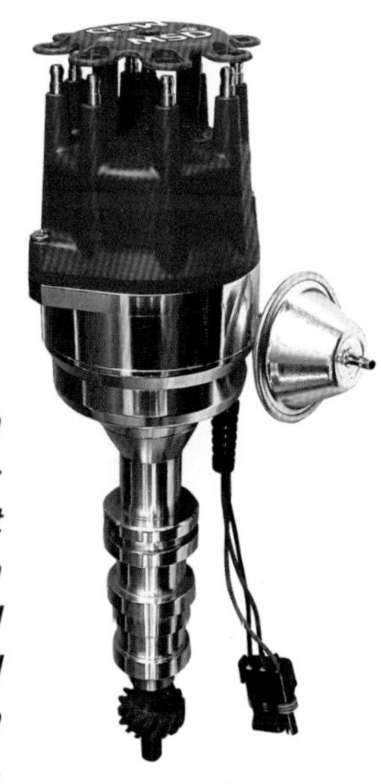

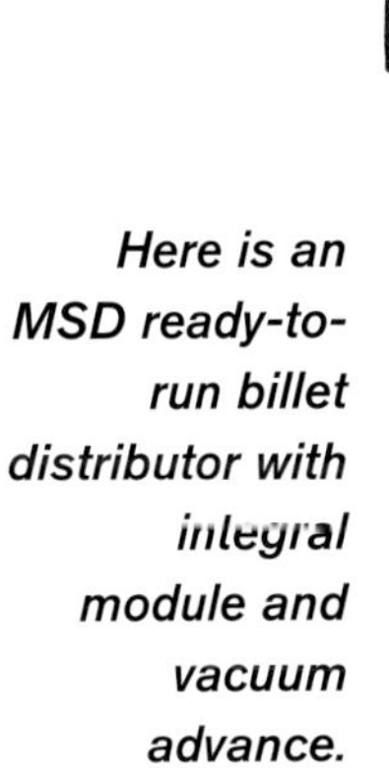

Here is an MSD ready-to-run billet distributor with integral module and vacuum advance.

I often use the 8.5-mm MSD wiring on race builds. Other manufacturers, such as Taylor or Moroso, offer comparable high-quality wiring products as well. Available in numerous colors, these wires are a good investment for any high-powered engine. But there are places where they just won't work out. A builder seeking an original appearance uses factory Ford-style valve covers, and the large-diameter wires do not fit through the plastic wire looms. Currently, I am using an MSD StreetFire universal plug-wire set on such projects. It is just small enough in diameter to fit the wire holders (barely) and the black color is unobtrusive.

Remember, when routing wires, you need to keep the plug wires for cylinders number-7 and -8 away from each other to prevent a cross-fire, which can really happen. Use dielectric grease on the boots to prevent seizing and voltage leakage to ground. Also, keep plug and coil wiring away from any electronic signal leads, such as the ignition trigger harness or EFI wiring. Inductive fields can play havoc and sometimes be nearly impossible to diagnose.

Spark Plugs

The factory FE cylinder heads accept the older-style 18-mm spark plugs with a 13/16-inch hex. The old Autolite/Motorcraft numbers were BF42—BF32 on the high-performance engines. Most new heads, including Edelbrock and Blue Thunder, use the more common 14-mm, 3/8-inch-reach plugs with a 5/8-inch hex. These smaller plugs are set deeper into the chamber, and reaching them can be a challenge because they're set into a machined "well" in the exhaust side of the head. Cutting down the outside diameter of a 5/8-inch plug socket on a lathe helps out considerably. I normally start my dyno tuning with a Champion Racing C61YC plug, but that is from my familiarity with reading them. Other similar plugs from other manufacturers certainly work as well. The Autolite 3924 seems to be a good selection for street-oriented builds.

Distributor Gears

The MSD billet distributor has a .531-inch-diameter shaft. If you need to change to a bronze distributor gear for a solid-roller cam, the gear from a 429/460 Ford works perfectly. Factory Ford distributors use a .437-inch shaft diameter, while the Mallory Unilite has a .500-inch shaft. All sizes are readily available.

The Ford and MSD distributors use a .125-inch-diameter spiral-wound pin for gear retention. In addition, they rely on a fairly snug press fit. If you are having problems with pins shearing or are trying to save a distributor with pin-hole damage, it is possible to drill out the pin hole diameter and use the larger-diameter roll pin from a small-block Chevy.

An FE distributor hold-down is now available in polished stainless steel. It has all the correct stamping contours, so it easily installs on the intake and keeps the distributor firmly locked into place.

Ignition Tuning

The FE tunes just like any other engine. Total advance depends on the compression, cam, available fuel quality, and combustion chamber and piston design.

In broad terms, an original-equipment iron-head engine wants between 38 and 40 degrees of total timing. An Edelbrock-headed package is best somewhere between 34 and 38 degrees total. A Blue Thunder head engine, with a more efficient chamber, wants between 28 and 36 degrees. But there is a huge amount of variability within those ranges, and every combination is different. The only way to tell what your engine really needs is to try it. As a caution, though, if you are really far away from those numbers, it's likely that something is wrong. A spun damper or a bad timing light can really throw you off course. And remember to correctly connect the MSD pickup coil wiring. I've seen that get reversed when going from the dyno to the car.

Idle-timing needs are dependent on idle-speed desires, transmission type, converter selection, and (of course) the chosen cam. You have two options if you find yourself with a car that idles perfectly and runs great but has too much base timing for the starter to spin it over. You can run very light springs in the distributor, which let you crank with less lead but bring in the timing as soon as the motor fires up. Or you can run an ignition bypass to spin it over without spark and then light it up once turning—cool in a race car but a bit awkward on the street.

MSD has discontinued its FE-specific crank trigger kit, but the mount can be fabricated. I normally use the lower two timing-cover bolts on the passenger side as the location for the hand-formed bracket. The rest of the parts (trigger wheel and transducer) are from the MSD small-block Chevy crank trigger kit.

FUEL SYSTEMS

The Holley 950-hp carb is like a 750 with an 850 base plate, and it's a great choice for a 500-plus horsepower FE. This carb atomizes fuel well and is an ideal choice for a smaller-displacement race piece as well as an aggressive street engine.

The fuel system used on the Ford FE engines was generally conventional throughout its years of production. An eccentric bolted to the front of the camshaft actuates a mechanical fuel pump, which is mounted to the driver side of the timing cover, and this supplies fuel to the carb. The vast majority of FE engines use a single carburetor (2- or 4-barrel) mounted to a cast-iron intake manifold.

Although produced in a limited number each year, the factory high-performance engines had a lot more variety in fuel delivery. Most of the high-performance engines used aluminum intakes (see Chapter 10). With rare exception, all the factory and aftermarket single 4-barrel intakes are designed to accommodate the Holley 4150 mounting pattern. Holleys were offered in a single 4-barrel, two 4-barrels, or three 2-barrels. The two 4-barrel engines used a pair of 4160 model Holley carbs mounted in line (more details later).

Ford carburetor packages (aftermarket and non-production that were available only through the parts department) were similar in configuration but also with a few notable exceptions. Those were the offset-mounted dual-quad setups used on the Edelbrock and Mickey Thompson cross-ram intakes, the early multiple 2-barrel intakes, and the Weber carburetor systems.

Mechanical fuel-injection systems, featuring eight individual runner stacks, have long been available for the FE in a variety of configurations. But these range in popularity from simply very unusual to extraordinarily rare. Current aftermarket electronic fuel-management systems are bringing this technology to the FE engine builder. Edelbrock now offers converted carburetor-style intakes with the fuel-rail mountings and injector bungs already in place on the Victor intake casting.

I'll touch on most of these, but the primary focus of this chapter is on readily available 4-barrel carburetor systems for street and mild race applications.

4-Barrel Carburetor Options

The vast majority of engine projects are topped off with either a single or a pair of 4-barrel carburetors. While this is not intended to be a carburetor book, I do need to cover the subject in some detail so you can get the most out of any engine build.

Autolite 4-Barrels

The non-performance engines from Ford often had an Autolite 4100-series carburetor. A surprisingly advanced design in a number of ways, the Autolite carb never developed a strong following as a high-performance part because, in part, Ford chose to run Holley carbs on all of the factory-issued high-powered combinations. That being the case, the Autolite has dual integral fuel bowls, easily changeable jetting, a low-leakage-potential bathtub design with very few gaskets below the fuel level, and annular boosters. While this carburetor is best used in a restoration-style project, where appearance outweighs power, it can still be tuned to run quite well. Ironically, they were the obvious design inspiration for the Holley 4010 and 4011 carbs of the 1990s, which have recently been re-released as Summit-branded carburetors.

Holley 4-Barrels: Basic Design Elements

The ubiquitous Holley 4-barrel is the indisputable carburetor of choice for the average FE builder. Holley 4-barrels come in a few different configurations, such as 4150, 4160, etc. Note that all of the technical information presented here also applies to the other carburetors using similar Holley-derived modular architecture.

Holley identifies its carburetors by model and then historically by airflow and design features. Holley carburetors and components have an engineering number cast or stamped in numerous locations. These numbers include an "R," such as 12R1234-5. The engineering numbers are useful for component identification, but the carburetor assembly's true part number is a "0-xxxx" stamped into the leading edge of the choke horn.

The standard-mounting-flange Holley carb is model 4160, with a metering block on the primary side and a metering plate on the secondary. The primary metering block has replaceable jets and a power valve. The metering plate is thin and has only drilled holes for fuel metering with no adjustability.

The Model 4150's carburetors have adjustable/replaceable jets and a metering block on both the primary and secondary side. They share the same mounting pattern as the less-expensive 4160, but are more readily tunable for various combinations.

Holley carbs, whether 4150 or 4160, can be either single or dual feed. A single-feed carburetor has a fuel inlet on the linkage side of the primary bowl, and a transfer tube running back to the secondary bowl on the opposite side. A dual-feed carburetor has so called "cathedral" fuel bowls that have an inverted "V" appearance as viewed from the end. They have a separate fuel line feeding each bowl—often joined together at a fuel fitting outside of the carb.

Model 4160 carburetors are almost always vacuum secondary designs, but manifold vacuum does not actuate the vacuum secondaries. They use a large diaphragm mounted to the side of the main body to open the rear barrels by sensing airflow through a bleed feeding into the primary venturi—assisted in some cases by another airflow-sensing feed on the secondary side. The signal generated by airflow past the bleed(s) is balanced against a spring in the diaphragm housing, which can be altered/selected to speed or slow the secondary opening rate.

The Carter carb and the similar Edelbrock version are street oriented and run well, but give up power to comparable Holley-style carbs. The clean barrel design of the Holley-style carb allows efficient fuel flow in wide-open-throttle conditions.

Model 4150 carbs can be either vacuum secondary or have mechanical secondary actuation. Only the dual-feed, mechanical-secondary carburetors are properly referred to as "double pumpers," which is a descriptive name taken from the necessary accelerator pumps on both primary and secondary sides of these carbs.

Dominator 4500-series carburetors are a completely different package because the main bodies and linkage are unique. However, they share external components, such as bowls with the traditional Holley 4150/4160 series. The Dominator was designed for racing from the very start and has an integral main and throttle body, along with replaceable boosters. The 4500 Dominator has a unique and large mounting flange and large throttle blades,

Quick Fuel Technology offers excellent race-style carbs with a lot of tuning flexibility, so the air/fuel mixture is correctly atomized at various throttle positions. The properly tuned engine is thus both powerful and responsive. This one is a 4500-series carb for larger-displacement engines and big power.

and they are all "double pumpers." Most of them carry airflow ratings of 1,050 or 1,150 cfm.

Airflow Ratings

Carburetor selection is most often based on potential airflow measured in cubic feet per minute. Subjecting each carburetor configuration to a test procedure as specified by the Society of Automotive Engineers (SAE) determined the airflow ratings. Unfortunately, the last decade has seen a drift away from this scientific process, however flawed, to a marketing-driven naming program in which the carb's assigned "SAE airflow number" now has no real value.

The SAE procedure rated 4-barrel carbs at 1.5 inches of vacuum drop. This meant that, while on the test fixture, cfm was measured at the point where airflow was high enough to generate 1.5 inches of vacuum below the carb's open throttle plates. At the rated test flow, the carburetor represents a restriction in airflow, which is not ideal from a performance perspective, but a fixed restriction value is necessary so the volume of air various carbs can flow is quantified. Hence, this provides a basis of comparison between carburetors. In essence, this means that the test is valid for comparison, but not definitive for selection by the rated cfm—a key flaw found in most of the common carburetor-selection formulas.

In addition, proper carburetor airflow testing was done on a "wet" bench. On the bench, a fluid with specified characteristics similar to gasoline was running through the carb's boosters and circuitry. Since fuel has both volume and mass, this has a significant impact on the amount of air that can flow through a given venturi.

Vendors of modified custom carburetors began to aggressively market their products, and it became common practice to test carbs with tweaked parts on low-capacity, dry-flow benches. These gave artificially inflated airflow values, which some consumers took to heart—making purchases without real knowledge of the actual gains achieved, if any. Eventually, Holley followed the marketing trend and assigned equally arbitrary numbers to its HP-series carbs. A Holley HP 950 is essentially a 750 main body with an 850 base plate, along with a host of other useful upgrades, but it is decidedly not a 900-cfm carb. A 1050 Dominator outflows a so-named 1,000-cfm 4150-style carb by a wide margin.

At this time, it is best to make your sizing decision based on throttle bore and venturi bore diameters. Discuss your project parameters with your chosen supplier, and listen carefully to their recommendations.

The single 1,050-cfm Dominator is the best choice for a 600-plus horsepower combination. Carbs are CFM size rated at 1.5 inches of vacuum drop below the throttle plates. This means that a carb operating at its rate flow is restricting air entry into the engine—not desirable in a race application. Bigger is better; I can make a big carb run well at part throttle with tuning, but cannot make a small carb "bigger."

Carburetor Boosters

In addition to the general airflow ratings procedure, there are a few other items that merit discussion before going into the selection process. First among these are boosters. The carb boosters are those small venturi rings that are sticking out into the airstream. The mounting legs are swedged/spun into position in the main body of the carb, and the small venturis serve to both increase the "signal" and to deliver the main-circuit fuel into the airstream.

Holley-style carburetors have one of three booster configurations, each with advantages. The least expensive is a simple straight-leg booster, which looks like a simple ring with a straight piece of tubing coming out of one side. The next design has a downleg booster. As the name implies, the downleg design "droops" from its mounting position and places the ring lower into the main venturi. They are more difficult (hence expensive) to manufacture, but also work better in most 4150/4160 applications—delivering more fuel with less of a flow restriction. The third is the annular booster; this design offers a larger cross section and an array of fuel-delivery holes, which are placed all around its diameter, as opposed to the single delivery orifice of the other two. Annular boosters are regarded as being more restrictive, but as starting full flow earlier in the RPM band and in a more evenly dispersed pattern.

Picking the Right Size Carb

There are a lot of various formulas floating around that use RPM, engine size, and assumed volumetric efficiency to select the right carburetor. As noted above, most of them are flawed because the carburetor airflow-rating process assumes a restriction.

A single 4-barrel downleg-booster-equipped 750- to 770-cfm Holley on a Performer RPM intake would be the best standard package for budget-oriented street use. The basic Performer intake is nothing special, but the Performer RPM is only a few dollars more, and offers the outstanding performance and response from idle all the way to wide open. Holley offers a wide range of tuning parts that are very easy to tune and optimize with only common hand tools.

I normally guess large on carbs, compared to most folks. I can tune good driving into a large carburetor but cannot make a small carb flow more air. On the 390-based 445-ci stroker motors, I've seen an increase of 20 hp on the dyno by going from a basic vacuum-secondary 750 carb to a downleg-booster double pumper with the same airflow rating. I've also seen a comparable gain going from that 750 double pumper to a larger 850. Although I am sure that the smaller carb might drive nicer, the bigger one always makes more power.

On the larger 482-inch FE stroker engines, I almost always use either a 2x4 setup or a single 4500 Dominator. The Dominator has proven to be a stronger package than any of the single 4150-style carbs I've tried, and the dual quads are better yet.

When choosing dual-quad carbs, there are a couple things to be aware of. On the factory-style FE 2x4 medium-riser setup, the two vacuum secondary 4160 carbs are mounted backward on the intake. The "secondary" goes toward the front of the engine and the linkage is on the passenger-side of the car. Model 4150 carbs do not fit. This unusual arrangement is for distributor clearance.

The factory 2x4 package has tremendous cosmetic appeal and performs quite well even on mild engines.

Here is a Holley 4160 factory 2x4 carb showing the correct side-linkage-pin-type location. An interesting aspect of the Holley 4160 is that it doesn't have a secondary metering block, which means that a short transfer tube flows fuel from the front to the rear of the float bowl.

During dyno testing, I observed that carbs with the secondary barrels and light springs in the diaphragm housings open pretty slowly, even on large engines. So I size the carbs as if they were a single double pumper and essentially pretend that the rear barrels are not even there—putting a pair of 750s on a street engine works just fine from a tuning and driving standpoint. If the engine requires the added airflow, the back barrels come in, and they can be tuned to some extent with lighter springs.

Power Valves

The power valves are vacuum-operated fuel-delivery devices that

This 2x4 driver-side view shows the linkage bellcrank that connects to the gas pedal. The connecting hose for the vacuum secondary canisters is also visible. It helps synchronize the secondary opening rate for the pair of carbs.

Here is an assembled 2x4 carb package showing the passenger side with the fuel log, as well as the S-shaped rear breather needed to clear the intake.

True race level projects receive fabricated components like this 2x4 SOHC intake with a pair of sideways-mounted Quick Fuel 850s.

This 2x4 throttle-linkage detail shows the proper orientation of the front-to-rear connecting bars. The idea is to set up the linkage so that both carburetors are at WOT at the same time. The linkage is be progressive. The front primary carb opens first, and the secondary carb opens afterward, but at a faster rate, catching up at WOT.

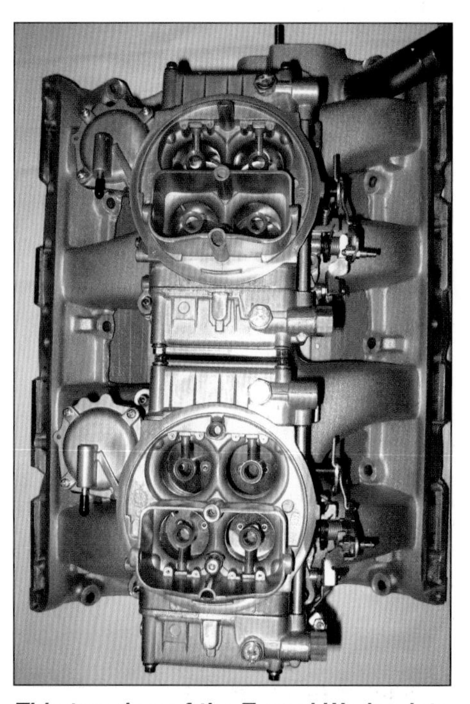

This top view of the Tunnel Wedge lets you see the carb layout and position relative to the distributor and rear openings. The MSD distributor cap sits higher than stock, and gets very close to hitting the front carb bowl.

work in concert with the main circuit of the carburetor to deliver the full amount of fuel necessary at wide-open throttle. The power valves are found on the primary metering block of almost all Holley carbs, and on the secondary block in many. Often improperly blamed for drivability issues, the power valves are an important factor in the total tuning package.

Power valves are normally held closed by manifold vacuum. They open and allow fuel to flow when vacuum drops below a predetermined level. The opening point for each power valve is marked on the body or washer face of the valve. A 6.5 valve is designed to open up at 6.5 inches of vacuum.

The power valves deliver fuel through two orifices or channels in the metering block called power valve channel restrictions (PVCRs). Fuel delivered through the PVCRs do not go through, nor is it affected by the main jets. Depending on the carb model, these channels measure from .032 to .093 inch in diameter and make up a considerable percentage of the overall fuel delivered. Power valves for Dominators may be different, though. Higher-flow models are colored gold and are needed with the larger-diameter channel restrictions.

A properly selected power valve and channel restriction combination allows leaner, and thus "clean and

This PCV passage can cause vacuum leaks if the intake does not seal it up. Check underneath with a dental mirror after carb installation.

This is another look at the Tunnel Wedge with traditional 4160 carbs in position.

crisp" part-throttle operation while providing the correct fuel delivery for WOT power.

A case can be made for removing the power valve from the secondary side of carbs, but you are going to have opened secondaries with manifold vacuum in many circumstances. Removing the power valve from the primary side is rarely the correct answer on anything other than a drag-race-only application, which runs at WOT only. In any case, removal of the power valve requires an increase in main-jet area that is commensurate with the area of the now non-functional PVCRs. You cannot simply assume some arbitrary jet-number increase. Each carb model has had a different

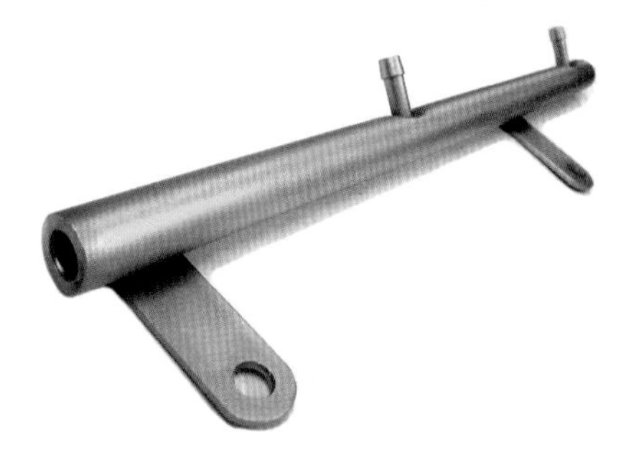

The factory-style fuel-distribution log mounts to two intake bolts, has a single hose connection for each carb, and feeds from the front.

The Carter is usually the best choice among performance mechanical fuel pumps. Some six-valve-style pumps don't clock the inlet and outlet separately, making line routing nearly impossible.

amount of fuel flowing though the power valves, and on some Dominator models it approaches 50 percent by area.

Blown power valves are not nearly as common as you might think. When bowl screws have been over tightened, the main body gets pulled in at the corners over time. The now-reduced clamping pressure at the center of the metering block gasket causes it to leak after a backfire—pulling fuel from the centered accelerator pump and power valve passages right into the vacuum cav-

TWM's electronic fuel injection system has the Weber look with modern performance. You still need to provide the electronic components to drive the system.

ity below them. The owner replaces the so-called blown power valve, sticks a nice new gasket in there, and fixes the problem until the next backfire.

The correct power valve is one that stays closed during light, part-throttle operation but opens when the throttle is "crowded" or under a fairly heavy load, such as when the vehicle climbs a hill. For example, if you feel a flat spot or get a surge under that driving condition with a 6.5-inch valve, you need a power valve that opens just a bit earlier with a higher vacuum number—maybe try a 7.5-inch valve. You also want a valve that stays closed at idle or at least is not pulsating at idle. On an automatic, you check vacuum in gear and then go an inch lower.

Floats, Needles and Seats

Holley carbs have a float bowl at each end of the carburetor. Although they at first look the same, the front and rear bowls are different. The float level on a Holley is externally adjustable, and verifiable with the brass sight plug on the side. The normally accepted standard is to have the fuel just at the bottom of the sight plug, but I prefer to set the level with the bowl removed and inverted, and the float parallel with the top of the bowl. I have seen cases where high fuel pressures allowed the fuel to appear okay at rest, but a low bowl setting lets the engine lean out at higher RPM due to inadequate float drop.

The plastic see-through sight plugs seem like a good idea, but I've seen them become hard and break.

The needle and seat have a pair of O-rings that must be in good shape. A cut ring either leaks externally or allows fuel to bypass the needle causing flooding. Old fuel or race gas with odd additives can cause the viton tip on the needle to erode prematurely.

There is a bumper spring that needs to be below the float arm when assembled—helping to pick it up.

Jets, Bleeds and Tuning

Tuning a carburetor seems pretty easy at first. Then it gets progressively more challenging as you learn more about the interrelationships between the circuits. This book is not intended to be a comprehensive tuning guide, but rather a series of discussion points to help you find your way. As a general rule, if you have to make big changes to get a carb to respond—like more than four jet sizes—you should likely have things checked out carefully before moving forward.

The best tuning tools are a good set of ears and eyes, followed closely by data in the form of oxygen sensors and brake specific fuel consumption (BSFC) numbers. In the first part of the tuning process, control is defined by a linear fuel curve. This provides air/fuel ratios that are reasonably smooth and respond predictably to changes. The BSFC numbers should track with the oxygen values, which means both show richer or leaner values together as changes are made. With control established, you can then make changes to see what the engine "wants" for peak performance at any given point in the RPM band or driving situation, and those bleed changes alter the power curve. If you see smoke or hurt plugs, or hear knocks or odd sounds, always recognize these problems and recheck the sensors and data before going further.

Main metering jets are easy to change and are the most frequently replaced tuning components in a carburetor. At least half of the time, they are changed for the wrong reasons, or are expected to cover for other issues.

Main jets work in concert with the fuel level, the emulsion bleeds in the metering block, the power valve channel restrictions, the boosters, and the main air bleeds. The overall fuel curve relies on the relationship between all of these, and a radical change in jet size alone can have some unpredicted results.

As an example, main jets are always "on," while the power valve circuit is only active at wider throttle openings. If you have good light throttle or cruising driving characteristics but are too lean at wide open, you should consider increasing the size of the PVCRs, rather than the jets. Increasing jet size fattens your part-throttle fuel more by percentages than it does the wide-open, and the PVCRs make up a good portion of wide-open fuel delivery.

The main air bleeds are at the top of the carburetor alongside the choke horn. These are sized to work with the booster design and the jetting/emulsion package to tailor the fuel-delivery curve and the onset of main-circuit fuel delivery. Changing the bleed diameter affects the starting point for the main circuit, the amount of fuel delivered by a given jet size, and the amount of air introduced into the main fuel stream. These changes can be pretty dramatic and a touch unpredictable at times due to the aeration factor. If you choose to do your tuning with bleeds, keep good notes, start with a small incremental change, and watch for trends before making any "big" changes.

Emulsion bleeds are the series of small holes drilled or installed vertically in the metering block main wells. The holes below the float level serve as added fuel feeds; those above act as additional air bleeds, and those in the middle as a combination of the two, which produces aerated fuel. As the engine RPM goes up, the fuel level normally drops and the overall bleed package changes. As with the main bleed, small changes can have a noticeable impact on fuel delivery. Plugging a bleed, or changing its size or position alters the overall fuel-delivery curve.

Accelerator Pumps

Holley-style carbs have a great deal of adjustment potential in the pump circuit. The squirter is the outlet for the pump, and it's found at the top of the venturi section. These are usually the first part changed, but they should be considered as part of a system rather than as a single item. Squirters are marked for flow, referencing the diameter of the outlet holes in them. Most carbs start off with something around .025 inch, but sizes of up to .037 inch are common for tuning.

The pump itself is below the fuel bowl, and can be either a 30- or 50-cc piece. The volume measurement referenced is actually an average of 10 strokes. The fuel from the pump goes from the bowl past an inlet check, and into the pump cavity. From there, the pump stroke pushes it through an upward-angled passage in the metering block, into the main body, and upward to the squirter mounting. There is a discharge check valve just below the squirter to prevent backflow and control fuel pullover through the squirter. The screw that holds the squirter in place needs to be a drilled piece for higher flow requirements of .035 inch or more.

A plastic cam mounted to the throttle lever operates the pump. The cam can be changed (many are available) to alter both the timing and volume of the pump shot. By changing the location of the cam relative to the throttle position, you can delay the onset of pump action, and this is useful for a drag car that leaves the starting line at an already-high throttle position.

The size of the pump cavity and the amount of stroke provided by the pump cam determine the amount of fuel available through the accelerator-pump circuit. Increasing the size of the squirter delivers the available fuel more quickly, and decreasing it slows fuel delivery. In vehicle testing, this is the only way to determine the correct amount of and timing of the pump circuit. A short hiccup followed by solid acceleration is a common sign of inadequate pump shot.

Fuel Injection Overview

"FE" and "fuel injection" are terms not often used together in the same sentence, but that is changing. The early mechanical fuel-injection systems that are found on eBay.com and in swap meets have a lot of cosmetic appeal, but they are not very street friendly when in good condition. And few if any of them remain in that shape.

On the other hand, I am now seeing an increase in the number of engines being retrofitted with electronic fuel-management systems—blending the cool cosmetics with modern functionality. These are most often using a carburetor-style intake manifold with injector bungs and fuel rails mounted above each runner. The throttle body takes the place of the centrally mounted carburetor, and incorporates a throttle position sensor (TPS) along with an idle air control (IAC). A 1-bar manifold air pressure (MAP) sensor is mounted in the engine compartment, while an inlet air temperature and a coolant temperature sensor are installed on the engine. One or two oxygen sensors are mounted into the exhaust system. The distributor, cam, and crank triggers are connected as well, to complete the connections required for the management system.

Engine management systems from F.A.S.T., Big Stuff 3, DFI, Holley, Edelbrock, and others are now available to control all parameters of engine operation. Each system has differing capabilities and requirements, but the similarities are enough to cover them as a group for this discussion. They generally control and monitor all fuel and ignition adjustments and have data-logging capability. Once everything is installed, connected, and programmed with the basic engine parameters, you should be able to fire the package up.

Using a laptop computer, you can set virtually any variable from timing and idle speed to fuel mixture without opening the hood. A pair of tables is used to define the basic setup. One reflects comparative efficiency of the engine at a particular load and RPM, and the other targets air/fuel ratio at those same points. It is a far quicker process to establish control of fuel delivery with the EFI system than it is with a carburetor. A tuner manipulates the efficiency tables until the corrections to a target air/fuel ratio are minimized. Once achieved, the same tuning process of finding what the engine "wants" is followed to get the desired results.

Choosing the proper injector size is important. Too large an injector has a slow duty cycle and does not deliver good low-RPM performance or idle quality. Two main reasons to go EFI in the first place is superior idle and low-RPM performance. A smaller-than-optimal injector can be crutched to an extent by using higher fuel pressure, and runs better on the street. A common rule is to target an 80-percent duty cycle at peak power, and therefore, 65-pound injectors support more than 800 hp at 45-psi fuel pressure—a set of 36-psi injectors supports more than 450 hp at that same pressure.

The benefits of an EFI package include better starting, driving behavior, and idle quality with wilder cams, because fuel delivery with EFI is not vacuum dependent. You can also get better control of fuel delivery in a wider range of operating conditions, such as under boost—a condition where a carburetor would have trouble compensating. The only real downside is the initial cost, which runs well into the thousands of dollars. Of course, race-quality carburetors are far from cheap.

EXHAUST SYSTEMS

The factory 427 1963 cast headers produce surprisingly good power. The passenger-side pipes can be seen here.

The 1963 driver-side factory 427 cast header is different than on the passenger side. Interestingly, there are slightly different ones available for 1965 cars as well, with revisions needed to clear the newer chassis design.

The exhaust system for the average high-performance and street-driven Ford FE engine consists of a set of tubular headers, running through a couple of mufflers, and perhaps tailpipes out to the rear of the vehicle—pretty common stuff. But there are plenty of available options within that simple description, and a few things to look out for. Some parts that should fit—don't. And some of the original cast-header-style manifolds are actually quite good.

Headers and Manifolds

The first modification everybody makes, when seeking more power, is adding a set of headers. And this is as true today as it was back when these cars were new. The factory cast-iron manifolds found on most FE-equipped vehicles can best be described as terrible, so headers make a very noticeable improvement.

What Fits—and Doesn't Fit

FE wedge cylinder heads all have one of three bolt patterns on the exhaust side. The vast majority of heads have an "up and down" pair of 3/8-16 holes centered vertically on each port—eight holes per head.

The Cobra Jet heads add an extra pair of "side-by-side" holes at each port opening for a total of 16. The vertical holes remain the same as more common castings.

The 390 GT heads have a unique 14-bolt pattern, with upper holes all across, lower holes only on the front and rear ports, and the side-by-side holes as found on the CJ heads. The upper holes on the 390 GT heads are not in the same position as on other FE heads either—making proper header fit a challenge.

All aftermarket heads have the traditional vertical bolt pattern, and many also include the Cobra Jet pattern, either as a standard feature or in optional head castings.

It is hard to believe that headers with as many as 16 fasteners could ever leak, but they sometimes do. And there is a reason.

Despite the fact that the vertical bolt pattern is the same on most heads, the location of the exhaust port relative to those vertical bolt holes changes from one head casting to another. Any FE header physically bolts to any cylinder head (with the possible exception of the 390 GT). But the exhaust port on some combinations are very close to or even overlap the header tube opening, causing a leak at the mounting flange. This is something that needs to be checked before installing your headers. The "fix" is normally a simple slotting of the bolt holes when necessary. You can also cut the flanges in between ports to allow for individual tube installation and adjustment.

Installing the fasteners in FE headers can range from simply difficult to nearly impossible. Header bolts with a reduced hex size are nearly mandatory—some are available with a 3/8-inch hex instead of the common 7/16-inch hex. Twelve-point headed fasteners seem like a good idea, but in many cases, it's difficult to torque them down with an open-end wrench. Be certain that your header bolts are not too long, or they will not easily go into position correctly, may cross thread, or even bottom out in the heads before getting tight. I have an assortment of inexpensive open-end and box-end wrenches that have been cut down and bent to ease this task. You'll also want a long 1/4-inch drive extension, appropriate swivel sockets, and a box of Band-Aids.

Header Selection

Despite the known fact that header design can have a significant impact on engine performance, you do not really have a lot of choices when selecting FE headers for traditional automotive applications. The most common are from Hooker Headers, which only offers a couple choices for each car, which are comparatively expensive.

The popular Mustangs and Fairlanes are limited to either a single 1.75-inch primary tube street header or a 2.125-inch tube multi-piece race

Here is a 1967 427 2x4 Fairlane engine. Note the unique exhaust manifolds, which were not used on any other application.

The C6AE-U 390 GT head with the exhaust side shown has a unique 14-bolt pattern. The upper and lower holes are not in the standard FE locations, so this makes the header fit problematic.

From top to bottom, the Cobra Jet, medium-riser, high-riser, and tunnel-port heads are shown from the exhaust side.

These Hooker race headers are multi-piece units that wrap around the motor mounts when installed in the car. Be sure your car has enough clearance where you're driving it because these headers are not speed-bump friendly.

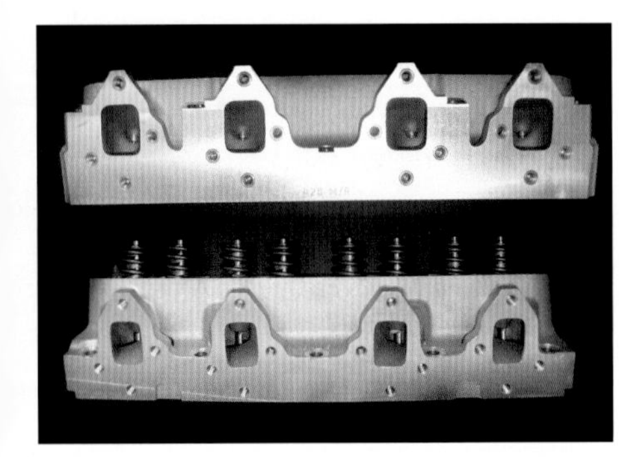

By comparing the Edelbrock and Blue Thunder heads' exhaust sides, you can see the exhaust port position difference relative to the head's deck surface. The Blue Thunder head exhaust is .400 inch higher, making header fitment more of a challenge, particularly in Mustangs or Fairlanes.

header. The street headers are offered as unique parts for the Mustang versus the mid-sized cars, but race headers are Mustang-only items that can be "hammer-fit" into a Fairlane chassis. Hooker used to sell a 2.00-inch primary tube race header that was a lot easier to package into the car, but it has since been discontinued.

Hedman also offers a limited selection of FE headers. In fact, there are few options, high costs, and challenging installation.

Ford Powertrain Applications (FPA) offers a variety of short-style headers that are quite a bit easier to install in many FE-powered vehicles. Given the challenges of installation, and the modest power difference between header styles in street-oriented applications, the short header is an attractive option.

Truck guys have it a lot easier on installation because there's plenty of engine compartment room, and the truck headers even perform better on the dyno. Headers for cars equipped with shock towers (like 1960s-vintage Mustangs) require the header tubes to make an immediate downward turn at the header-mounting flange, and this adversely affects flow. On truck applications, there is enough room for a straight section coming out of the port before any bends are needed.

When building a package for a +/-500-hp FE, the 1.750-inch primary tube headers seem to be perfectly adequate. With 600 hp or more, you start to see benefits from a larger 2.000-inch pipe diameter. I have seen nearly 800 hp with the 2.000-inch Hooker headers on the dyno. While there is certainly a point where an even bigger tube is needed, the headers need to fit into the chassis. Engines producing 800 hp go into a race package, and in these cases, fabricating custom headers is part of the budget. Therefore, the issue is moot.

On the dyno, I have seen significant changes in torque from header extensions. If you are running with open exhaust, you would be startled to see just how much you can gain—or lose—with a change in collector length. Collector length seems to be more important than primary pipe length by a significant amount, based on my dyno experience.

Cast Manifolds

As mentioned earlier, most factory-cast FE exhaust manifolds are pretty restrictive, but there is a very

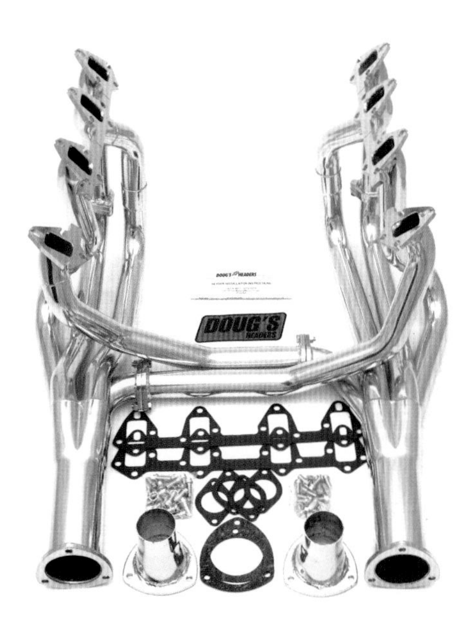

These recently re-released Doug Thorley race headers criss-cross under the engine. I've never tested them, but they certainly look impressive.

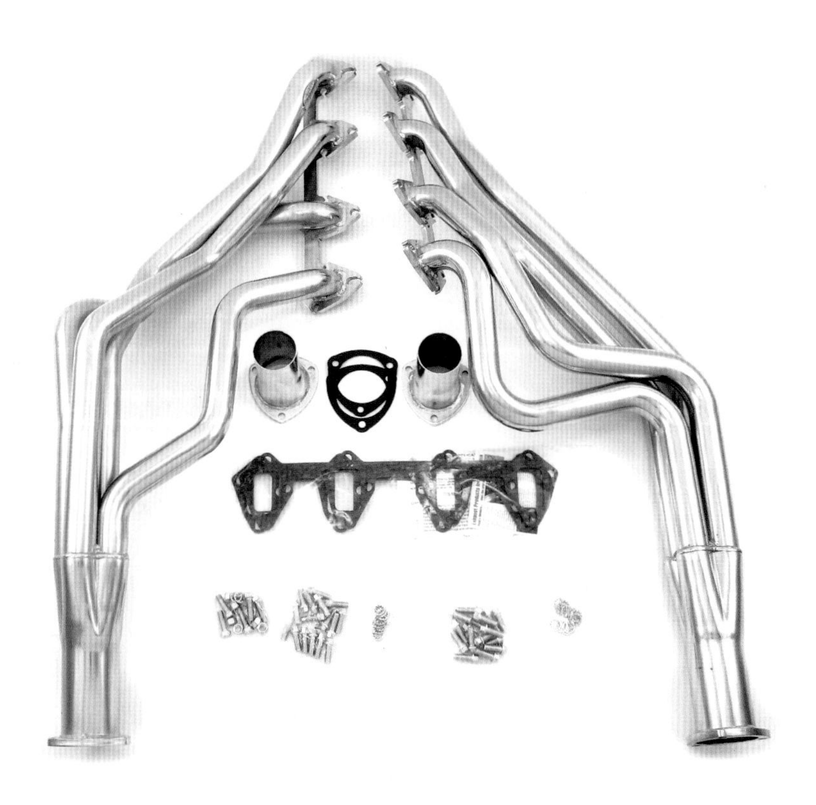

These street headers have slotted mounting holes, which helps ease installation. However, you still need to test fit them for port alignment before mounting them.

notable exception. The cast-header-style manifolds on 1963–1965 427 Galaxies are startlingly good parts. The designer of these knew what he was doing some 40 years ago. I have tested the cast long-style manifolds on an engine that made 636 hp with Hooker 2.000-inch race headers and only lost 30 hp. If you are building a Galaxie, and can afford the approximately thousand-dollar price-tag, the factory-cast long-tube manifolds are a sure winner. They fit, they're quiet, they don't leak, and they run very, very well.

Custom Headers

Hand-made custom headers definitely get the most potential power from your engine combination. A header fabricator can optimize pipe sizing, length, and collector designs around any performance parameter. There is a lot of science surrounding header design, but the FE faces real packaging challenges that effectively reduce the potential benefits.

Getting the "right" pipes to fit into the tight confines of a Mustang is a challenge for even a skilled fabricator. Fabricators work on a pay-per-hour basis. You get better performance, but also a significantly lighter wallet.

Mufflers

Muffler selection is a very personal decision. Everybody has a different idea of what is "too loud" or "too quiet." In addition, some combinations exhibit drone or resonance at certain speeds, which is something that cannot be predicted. In my dyno- testing experience, I've had pretty good power results from the popular straight-through designs from Magnaflow, but they are loud. I've also done quite well with the much quieter Hooker AeroChamber muffler, which has an unusual internal design.

Gaskets

I have had the best luck using header gaskets from Fel-Pro. These have a perforated steel core, which seems to hold up very well. The inexpensive non-reinforced gaskets do not stay in one piece for very long. Since it's tough to change header gaskets in an FE, you want to use the best possible parts. I have also had good results from using a very thin layer of high-temperature copper RTV silicone on both sides of the gasket—just enough to seal up any irregularities in the gasket surfaces.

EXTERNAL ACCESSORIES

The ATI Performance Products damper is likely the best part in the marketplace, but it requires minor component modifications in order to work. The timing location is off from the factory FE position, and the pulley bolt pattern is big-block Chevy.

With any engine project, a large collection of external components takes up a fair portion of the engine build time and budget, yet many are not considered parts of the basic engine package. Among these are the bolt-on items, such as valve covers, motor mounts, water pumps, and breather systems. This chapter gives an outline of what is available and what is needed to complete your FE build, along with a few tips on what challenges to look out for.

Motor Mounts

This was touched on in Chapter 2, but it merits repeating. The FE engine has had two different motor-mount bolt patterns over its production history. The earlier engines had a 2-bolt mount pattern while later (post 1964) engines had a 4-bolt pattern cast and drilled into the sides of the block. The later-model blocks can be retrofitted into earlier applications because the two original bolt positions were retained. But early blocks cannot be used in later-model vehicles without fabrication. Later-model vehicles often use only two of the fastener locations, but not the same two as the earlier vehicles.

Also, 427 side-oiler blocks, along with the available aftermarket replacement blocks, require a modified engine mount on the driver side in order to clear the oil-feed casting rib. When installed, the FE motor mount isolators sometimes appear "wrong" because they attach to the engine on an angle, not parallel to the oil-pan rail.

Race cars often use a front motor plate that attaches at the water-pump mounting holes. An attractive and perhaps stronger alternative would be the marine-style mounting for timing covers, which incorporate a flat machined surface on the lower front.

Expansion Plugs

The common 360/390/428 FE engine uses six formed-metal expansion plugs to seal off the core openings in the side of each block casting. These *are not* freeze or frost plugs, and do not prevent block breakage. They are simply openings to allow the casting sand from the inner cores to be removed during manufacturing.

The expansion plugs are simply hammered into place with anaerobic sealant around their perimeter. It is important to note that the FE block requires an odd-sized 49/64-inch-diameter expansion plug. Some plug kits provide a cheaper 3/4-inch plug, which does not seal effectively and will possibly come loose!

The earliest 427 engines also used press-in-type plugs, but the later ones went to screw-in-type core plugs for added strength and reliability. The factory screw-in core plugs are a shallow design with a tapered pipe thread. The plumbing-style plugs you see at the local hardware store are too thick/deep and do not go in flush to the sides of the block. Passenger car 427 engines used steel plugs, while the marine versions used brass. Since the plugs are nearly impossible to remove after more than 30 years of service, the brass plugs serve as a clear indication of a marine engine.

This is a typical FE brass-core plug kit. The same plugs fit the vast majority of engines. The only cautions are to make certain they fit tightly, and to remember that the cam plug goes in with the flat surface facing out.

FE cam plugs install with the open end facing toward the cam—backward in orientation compared to a core plug.

You absolutely need to use the correct automotive screw-in core plugs to ensure optimal fit and prevent potential engine damage. The correct screw-in core plugs have a shallow depth while generic hardware-store core plugs are often too thick and do not properly seat, or interfere with motor mounts.

Genesis aftermarket blocks feature a reproduction of the factory screw-in plug that provides a good seal, while the Pond block uses a CNC-formed plug with straight threads and an O-ring for sealing. It is possible to convert a normal FE block for screw-in-style plugs, but it is not an easy task. You need at least two thread taps—one to get started and another shortened one for finish-sizing the threads. The distance between the outside of the block and the cylinder cores is too short to allow use of a standard pipe tap. You *do not* want to run the tap into the cylinder core.

Block Plates

Early FE engines did not use a plate between the engine block and the transmission, but they were required on most of the applications you are likely to find in a muscle-car-era 360, 390, or 428 block. These thin metal plates seem innocent enough, but can cause an occasional problem.

If you are experiencing starter troubles, it's worth noting that the FE starter mounts to the transmission and not to the block. This means that the block plate, or lack of same, can have an impact on starter drive-gear engagement.

The other block-plate issues are related to the various galleys and core plugs in the bellhousing end of some blocks. The threaded-in oil galley plugs surrounding the camshaft often protrude from the block surface enough to interfere with the plate. Many factory block plates already have clearance holes for those plugs. If your block plate does not have them, they are easily added. You can test fit the block plate by tapping it with a hammer where the plugs touch, to mark the position. Then drill the appropriate holes. I use a Greenlee punch from the electrical supply house to make clean, round holes.

The Genesis aftermarket 427 blocks can have a similar interference issue with the large core plugs at the rear of the block. The fix is the same as for the galley plugs; locate the needed hole and use a Greenlee punch (or hole saw) to provide clearance.

Some FE engine-block plates like this one have the extra holes needed for oil galley plug clearance, while other plates may need to have them added. Very early engines did not use the block plate. Although it may be unlikely, the block plate can be the issue if you have a difficult-to-diagnose starter.

Water Pumps

The common FE engines all used a similar-style water pump, which straddles the timing cover and is mounted to the block with four 3/8-16 bolts. The bolts extend into the water jacket and require sealant to prevent leaks. The inlet hose for the pump is on the driver-side lower corner. A bypass fitting is centered in the top of the pump that connects to a short piece of 5/8-inch ID hose, which runs to the front face of the intake manifold. There is also a fitting for a similar-sized heater-hose connection on the upper passenger side. It is common for both the heater and bypass fittings to be plugged on race applications. If running without a bypass hose, you must also either drill bypass passages into the thermostat or run without a thermostat (not recommended for street cars).

Along with the rebuilt water pumps readily available "at parts stores," the FE builder can choose from a few new aftermarket parts. Upgraded cast-iron water pumps are available from a few suppliers including Milodon. Aluminum pumps for weight-conscious projects are available from both Dove and Edelbrock. Both claim enhanced flow capability for better cooling; a feature that has definite benefits in the tightly cramped engine compartments common to FE installations.

Electric water pumps are also available for the FE. I've used the parts from Meziere with excellent results even in street applications. While less effective at delivering coolant pressure and flow at higher speeds, the electric pump has advantages in drag-strip environments because it permits between-rounds cooling, and allows for packaging flexibility. The electric pump also frees up a modest amount of horsepower in a race engine. The use of an electric water pump requires some creativity in alternator- and accessory-drive fabrication.

Front Accessory Drive Systems

The front of FE engines has any of a vast array of front-end-drive systems. These range from a simple single-pulley alternator (or generator) belt to multiple-belt arrangements driving air conditioning, power steering, and emissions air pumps in various positions.

The most common and simple arrangement is a single, long pivot bolt that is installed into the 7/16-14 threaded hole on the passenger-side front of the block. The bolt goes through a triangular steel bracket in the front of the alternator and through a spacer that is placed behind the alternator. The other portion of the bracket is mounted to the water pump. The sliding mount and bolt used for belt-tension adjustment attaches to a longer, formed-steel bracket, which also mounts to the water pump. The alternator may be positioned either above or below the pivot bolt depending on application.

Earlier-model engines with generators used a bracket that mounted to a single fastener along the side of the block. In later applications, that fastener hole was retained, but used for the ground cable instead.

Power-steering pumps, when used, are mounted to the driver-side cylinder head. The most common steering-pump layout has a cast-aluminum mounting bracket that bolts to the pump.

Pulley kits are better looking than the stock pulleys and are an ideal choice for high-performance street and show engines. These kits are available from several suppliers; this one is from Billet Specialties.

Air conditioning and/or emissions air pumps use a variety of mountings and idler pulleys that are entirely vehicle dependant. The system from a Galaxie may not physically fit into a pickup truck or a Fairlane. If you are replicating an original system, be certain that you acquire everything from the donor vehicle or the supplier. Mixing and matching factory pulleys can get you into trouble because they often have varied positions relative to the mounting. Therefore, the double pulleys on a crankshaft don't line up

This factory steel damper spacer goes between the timing cover and the damper. These often get worn or damaged over time. They should install with a light push fit. A muffler clamp and a three-jaw puller can be used to remove stubborn spacers.

Damper spacers with the balance "hatchet" for the 428 SCJ are also available as new items from Blue Thunder, and they are required to achieve engine balance when SCJ rods are used.

A factory single-groove lower pulley is shown here. It has mounting holes opened to fit the ATI damper's big-block Chevy bolt pattern.

Aftermarket FE damper spacers are available in either anodized aluminum or steel. After 40 years of service most original ones have a groove worn into them from continuous contact with the front seal, causing oil leakage.

A factory damper with pulley is often badly weathered. Replacement or reconditioning is highly recommended.

This is an OEM damper bolt and washer. Note that the washer needs to be very thick in order to handle the bolt torque load. Thin washers, even when stacked, deform and allow the bolt to loosen.

to a single on the water pump and small-block or 460 parts do not freely interchange.

Aftermarket belt-drive systems are available from several suppliers and range from simple aluminum replacement pulleys to fairly elaborate serpentine-drive combinations. As when working with factory parts, it is important to verify belt alignment on the aftermarket systems. The variance in pulley-mounting position on the aftermarket dampers can cause issues if not addressed.

Either the original or the aftermarket crankshaft pulley need the mounting pattern altered when using an ATI damper. The ATI damper uses a Chevrolet mounting pattern, which is slightly different than the factory FE bolt circle.

If you are using a crank trigger ignition, you need to shim/space all the other accessories forward 1/4-inch to compensate for the thickness of the trigger wheel.

The ARP damper bolt is available in either 12-point or square-drive recess for a 1/2-inch ratchet.

Dipstick

The oil-level dipstick on the common FE is located at the forward end of the block, where it is pressed into an opening just behind the oil-filter mount. Reproduction sticks and tubes are available. I often simply use a piece of 3/8-inch fuel line tubing bent for header clearance and cut to the appropriate length, with an upper clamp attached to a convenient bolt on the head. The stick itself can be a common 302 replacement with the "full" position marked as required.

When using a windage tray, you sometimes find that the dipstick hits the tray when inserted and curls back upward into contact with the crankshaft—not good. Usually you would check and address this issue during assembly. If you cannot simply bend the dipstick and feel your way through/around the obstacle, then you must make an adjustment. Just remember to do so every time you check the oil, or you may have a stick break off and send debris through the engine.

Many aftermarket "T"-type oil pans have provisions for a dipstick going into the top of the kickout. If you choose to use a stick at that location you'd normally plug the original location with a small piece of aluminum rod tapped into place. There is no functional advantage of one location over the other; it's simply a matter of convenience.

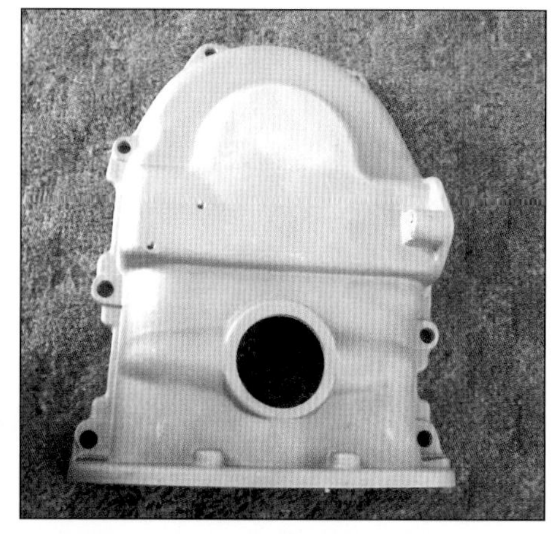

Genesis sand-cast timing cover is functional but heavier than the stock parts.

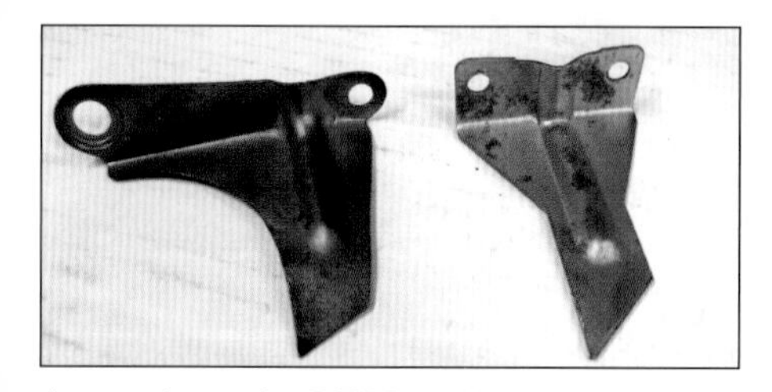

An early-style OEM timing pointer is on the right and a late-model style is on the left. The timing location is the same, and the covers can be interchanged as long as the matching pointer is used.

Valve Covers

Lots of valve-cover options are out there for the FE project. Choices include original tin in painted or chrome finishes, factory aluminum castings, and a plethora of aftermarket parts. For those builders looking for an original external appearance, it's good that all FE valve covers (with the notable exception of the Cammer) share the same mounting pattern. But some of them do not clear certain valvetrain combinations without modification.

The pre-1965 factory tin covers had a rounded appearance with no oil-fill or breather openings. Some of them had brand or model designation lettering stamped into them, such as "THUNDERBIRD," while others were completely smooth, which the FE-building community refers to as "baldies." Chromed baldy covers were used on the 1963 and 1964 427 engines, and were adorned with the famed "427 logo," a gold-and-black decal; a stylized bird flanked by checkered flags. The baldy-style covers have no internal baffles and seem to clear all valvetrain systems—original or aftermarket. In order to use them you need to provide oil-fill and breather openings

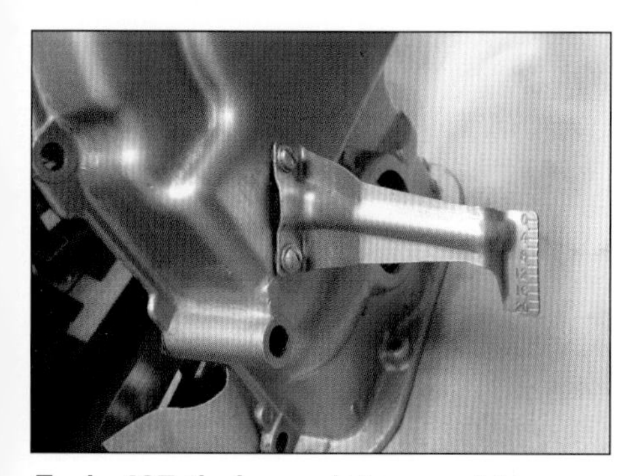

Early 427 timing pointer must be used with a matching "no numbers" damper.

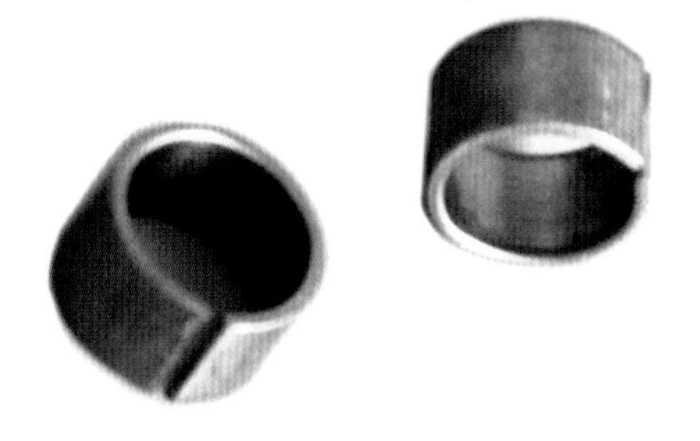

FE head dowels are simple split-and-wound pieces. These are the same as those for a 460 and are readily available.

The 427 "bird-and-flags" logo is normally used on chromed, stamped-steel "baldy" valve covers—T-bolt style.

This is a Mercury pentroof valve cover; similar Ford covers have no logo stamped onto them.

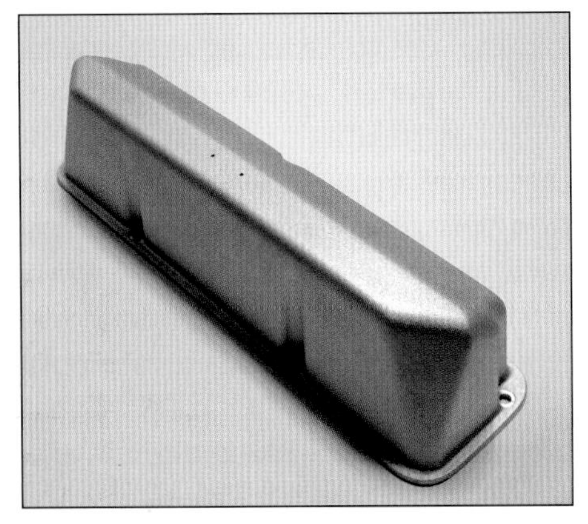

Blue Thunder cast FE pentroof valve covers share the factory look, but these aluminum castings seal better than the original pieces.

in the intake manifold. The chromed baldy covers and the decals are being reproduced.

The 1965 Ford 427 engines moved to a tall, sharp-cornered valve cover that is referred to as a "pentroof" design. These are also found in chrome and painted finishes. Later variations had oil-filler and PCV/breather openings and baffles in the cover. The tall pentroof covers usually accommodate most rocker combinations. A much shorter version of the pentroof was common in non-performance applications. These shorter covers do not provide the needed room for many rocker systems without removing the baffles and sometimes even that is not enough. The chromed tall pentroof covers are being reproduced either with or without filler openings.

In 1968 the FE engines received yet another valve-cover design. These are the most common "Powered by Ford" stampings that carried on through the end of production. Some of the performance vehicles (390 GT and 428 CJ cars) had the covers chromed, but the vast majority of them are painted. They all have oil fillers, breather/PCV openings, and internal baffles. They provide clearance on many aftermarket rocker systems, but not always. Usually, removing the baffles gets them on okay but sometimes a double gasket or a valve cover spacer is needed. Chromed "Powered by Ford" valve covers are being reproduced.

Factory aluminum covers are also readily available, either as swap meet finds or as reproductions. Blue Thunder is the aftermarket supplier of choice for most of these, and it has a wide array of cast pentroof-style aluminum covers. Most popular would be the classic finned cover with the "Cobra 427" or "Cobra Lemans" logo. These are also sold as a plain finned cover with no logo and as a non-finned, smooth-finish, aluminum pentroof. These all have valvetrain clearance for the most popular systems but may require clearance grinding for use with large-diameter valvesprings.

The 428 CJ/SCJ engines also had a factory die-cast-aluminum valve-cover option. Available as reproductions, these share similar fitment with their "Powered by Ford" steel siblings, usually okay but always need to be checked. The finned CJ covers are available as reproductions.

A Blue Thunder plain-finned valve cover set has a different look, which is closer to the 427 Cobra covers in profile but without any logo.

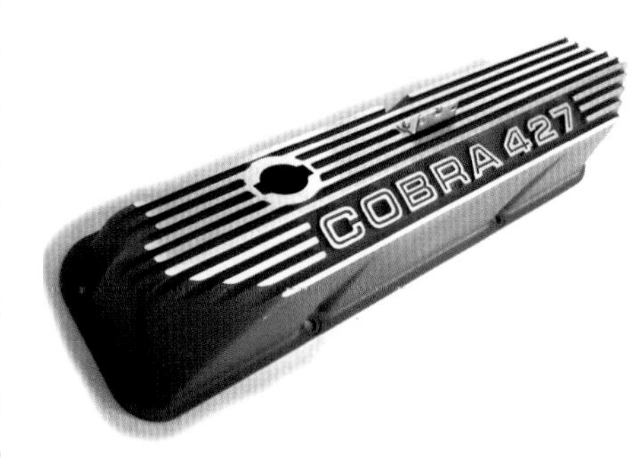

Cobra 427 valve covers are something of a classic visual cue on an FE.

The Blue Thunder Competition Style FE valve covers clear most any valvetrain, and have a more modern appearance.

Non-reproduction aftermarket valve covers are commonly the popular squared-off-finned-style designs from M/T, Cal Custom, or Edelbrock. Similar to the ones that each

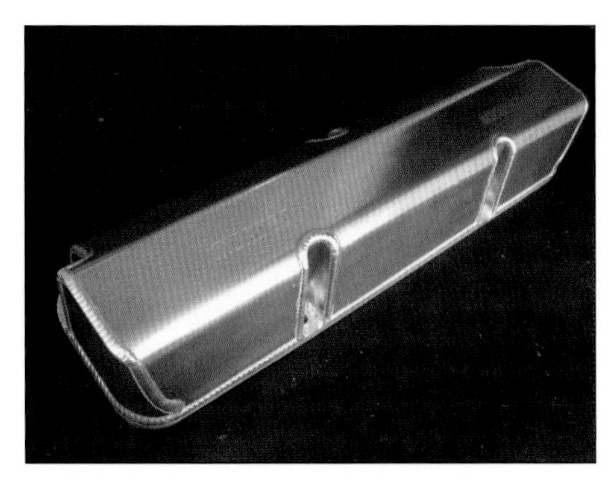

Moroso sheetmetal valve covers provide a hard-core race cosmetic, and they're both rigid and fairly lightweight compared to a cast cover.

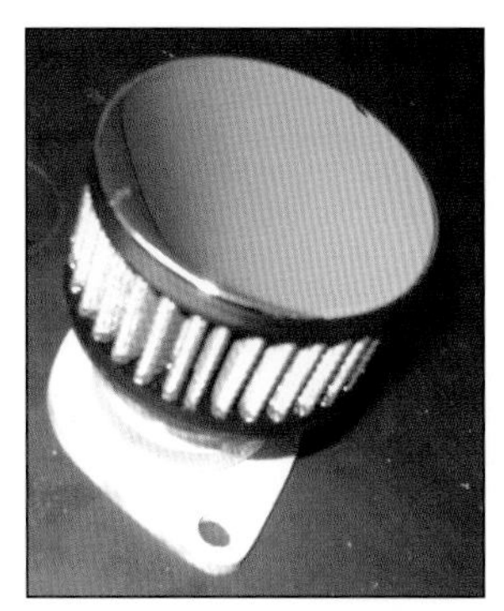

This fabricated breather kit is used for added crankcase ventilation when the valve covers have no openings. These are fabricated by Survival Motorsports by welding a threaded breather bung into a machined-aluminum rear cover plate.

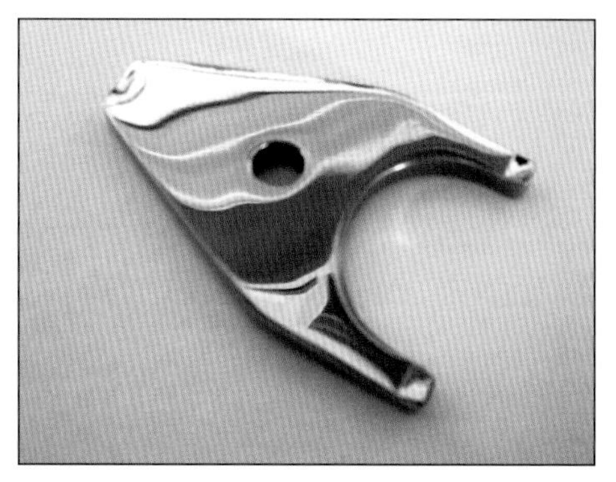

Distributor hold-downs are now available in either polished or brushed-finish stainless.

company sells for other engines, they are used most often in vintage-appearing hot-rod builds. Most Cobra and Mustang owners either opt for the factory look or the modern Blue Thunder "Competition Style" covers. The Competition Style covers are a visual departure from the normal FE design, with sides that extend to the edge of the sealing surface and inset hex-headed fasteners that reach to the top of the cover surface. In addition, these covers have a threaded bung in each end to accept a billet breather, oil filler, or PCV cap. They clear any valvetrain system.

Breathers, Oil Fillers and PCV Systems

Every engine needs a couple breather openings, with the exception of a pure race engine that runs a crankcase vacuum pump. Without adequate ventilation, even the best engine's crankcase would become pressurized due to blowby, and would suffer from serious leakage and moisture contamination.

Earlier FE engines had no openings in the valve covers and relied on the oil-filler cap/breather at the front of the intake manifold along with a road draft tube at the rear. The road draft tube assembly used a basket with woven mesh in it as a liquid separator. This basket sat in an opening in the rear of the intake, and the blowby gases were simply released into the atmosphere. Air passing the angled end of a tube that extended down toward the road surface generates a degree of suction. For builders wishing to retain the original-style valve covers but not wanting to run a road draft tube, I've fabricated a rear breather that uses a small filter on top of the basket—it's unobtrusive and seems to work well.

In 1966, Ford installed positive crankcase ventilation (PCV), and the breather openings were moved to the valve covers. Placing the breather system at a higher location on the engine is arguably better, but cosmetics sometimes drive the placement as much as function. On a PCV system, you have a valve on one cover and an inlet filter/cap on the other. Vacuum at idle or cruise pulls residual crankcase vapors into the intake where they are burned along with the normal fuel/air mixture. Good for emissions and air quality, but not ideal for performance. Most race-style applications do not use a PCV system, but they are not at all detrimental for street use, and may reduce tendencies for external oil mist and leakage.

For many years, serious race engines simply used one or more filter breathers on each valve cover to vent blowby to the atmosphere. Many still do. An open breather system works well, but you should expect to see some mist or oil film from the openings after a period of time–no engine completely eliminates blowby. A racer would not be concerned with the need for an occasional wipe from a shop towel, but a show car may be well served with a cleaner PCV-type system.

Over the last decade, dedicated racers have moved to vacuum systems. Using a belt-driven pump, race engines pull a 10- to 15-inch vacuum in the crankcase. A dry-sump oil pump or a dedicated vacuum pump can generate vacuum. Any excess oil vapor or mist is accumulated in a tank plumbed to the outlet side of the vacuum pump. This system allows the use of a lower-tension piston-ring package and is said to deliver significant horsepower gains (20 hp or more) when used in an optimized engine. In street applications, vacuum systems are marginal if usable

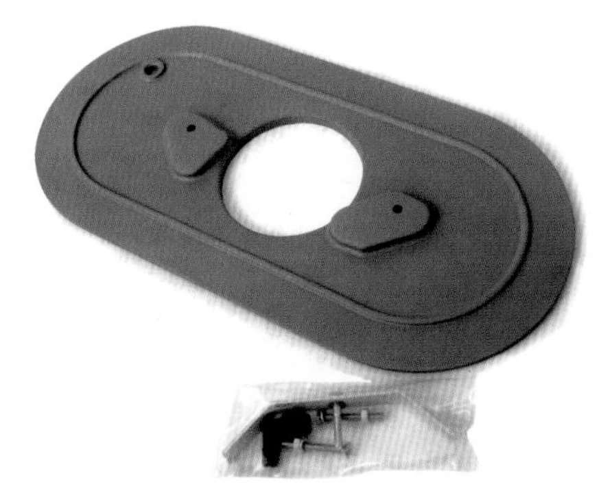

The single 4-barrel oval-style air cleaner base lets single-carb guys run the classic oval air cleaner that looks like a dual-quad air cleaner.

The 2x4 air filter elements are available in 1.75-, 2.00-, 3.00-, and 4.00-inch heights.

The air cleaner base for three 2-barrel carbs has D-shaped openings.

The air cleaner lid for the 3x2 setup is unique in both bolt spacing and fin design.

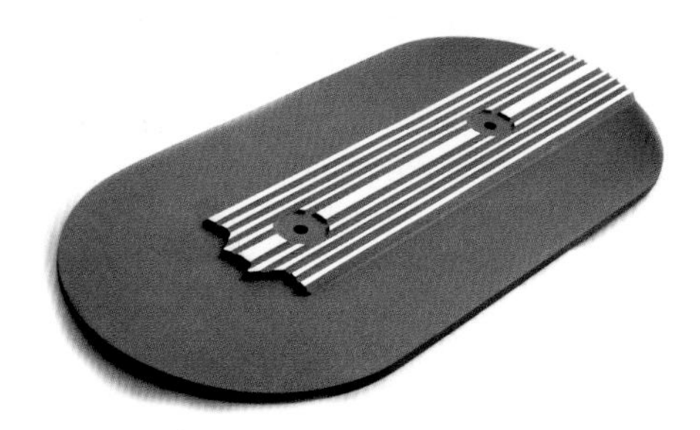

The popular 2x4 air-cleaner lid is sold with a relieved area forward of the fins to accommodate the "bird and flags" appliqué.

at all. In order to function, the engine needs to be extremely well sealed up, often requiring specialized or reversed front and rear crankcase seals.

Air Cleaners

The common 4x3-inch open-element air cleaner with a chrome lid works great and looks perfectly fine on an FE. But there are a lot of other alternatives as well, especially for those desiring a period-correct appearance.

The cast-aluminum, finned, oval air-filter assembly is instantly identifiable as a trademark FE item. This filter housing was originally installed on 390, 406, and 427 engines with the 2x4 or 3x2 carburetor packages and it is nicely reproduced by Blue Thunder. It also serve as inspiration for similar designs from numerous manufacturers. The lids are available with the inset for the "bird with crossed flags" emblem or with full-length fins. Bases are offered for the 2x4 and the 3x2, as well as for a single 4-barrel. Elements for the oval air-cleaner assembly are still available in paper filter media, and are also available from Blue Thunder in oiled fabric in heights of 1.75, 2.25, 3.00, or 4.00 inches.

Those wishing to duplicate the appearance of a Thunderbolt or Galaxie lightweight will be searching for the aluminum air box that connects via convoluted hose to the inner headlights or grill. It's an elegant aluminum casting but holds no filter—strictly a race-car part. Beyond unfiltered air, the only caution is that the original high-riser intake had different carburetor center-to-center spacing compared to medium-riser intakes. Most aftermarket intakes use the medium-riser spacing. Make certain that your air box matches your intake; both spacing options are available.

Among the other factory air-cleaner assemblies, the optional shaker hood assembly found on 1969–1970 Mustangs is the most notable. You can use a shaker hood when running a Performer RPM or a Blue Thunder intake manifold, although the position relative to the hood opening is a bit different with the RPM.

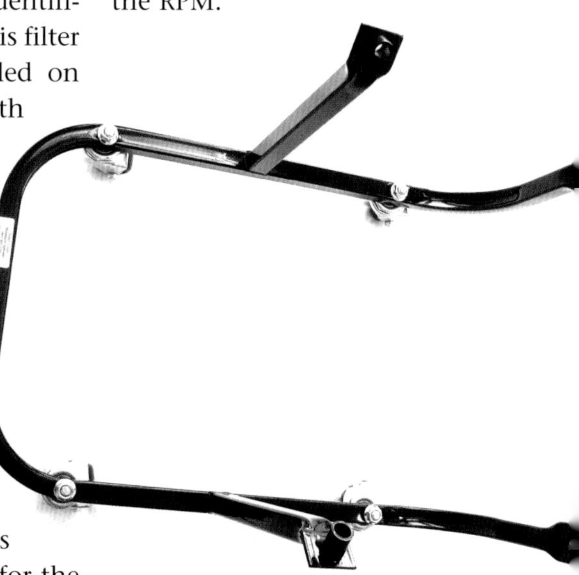

Engine cradles allow simplified engine transport and storage.

FE ENGINE ASSEMBLY PROCESS AND TIPS

Link-bar-to-block interference occurs on some Genesis blocks. It's much easier to find and fix it now, rather than later when you're trying to locate that strange tapping noise.

This chapter is not just a basic step-by-step section for "how to assemble your engine." I assume that you already have a basic understanding of engine building and assembly procedures and will focus on those things that make an Ford FE project unique, along with procedures that will improve your engine's performance. The information is organized in engine-build-process order. Some of these concepts have been addressed in earlier chapters, but warrant repeating in the context of the overall build process.

Finishing the Block Prep

With your chosen block all finish-machined and mounted to an engine stand, you can prep for assembly. All desired grinding and oiling modifications have been made. The next thing to do is check every threaded opening and oil passage in the block one last time. Few things are more frustrating than having to disassemble a block for more drilling, cutting, or grinding.

For final cleaning, I use a variety of products, one after the other. The first is hot water with a strong detergent. Using rifle brushes and household scrubbing brushes, aggressively go over every surface and follow with a stream of hot water from a garden hose. Mounting a Moroso cylinder-cleaning brush into a cordless drill seems to do a really good job of cleaning up the cylinders.

Certain types of debris seem resistant to hot soapy water but readily come off with brake cleaner. WD-40 removes yet another layer of crud, such as machining oils and embedded debris. It is always a surprise to see how much grime and debris remains behind after the first wash. Once confident in the cleanliness, I use a rubber-tipped air-blow gun to thoroughly dry the block, concentrating on all the threaded holes and galleys. The last step is to wipe down all machined surfaces with WD-40. If a machined surface is truly clean, it will rust almost instantly.

Make a last check for missing galley plugs. The one behind the distributor is easy to miss, as are the small ones used on the oil-pan rail on original side oilers. Also look out for the 1/8-inch NPT plug near the oil filter mount on some truck and service blocks because it was an accessory compressor oil-return line provision.

Cam Goes In First

The first parts installed in a new engine are the cam bearings. I've used a fabricated tool to pull in the bearings, but the common hammer and expanded mandrel installer works just fine as long as you use it properly and pay close attention. Installing the front cam bearing from the rear of the block allows use of the alignment cone, giving you a better chance of getting it in straight.

On a traditional top-/center-oiling FE, the location of the oil-feed hole in the center three cam bearings is not critical due to the annular grooves in the block. Most builders install them at the 3 o'clock position. The front and rear bearings only go in one way to line up the feed holes. On side oilers, the cylinder deck feeds must line up with the holes in the cam bearing. You can darken the build room and use a small LED flashlight to look directly through the hole you're trying to line up.

With cam bearings in place, I always test fit a known straight and dimensionally correct cam to verify the installation. Bearing deformation during installation causes most binding and tight spots. A careful effort with a knife and a Scotchbrite pad often remedies the problems. New cams can be bent, or have journals on the high side of diameter specs. I've turned an old camshaft into a bearing cutter by machining a flat onto each journal, leaving the corners sharp, which is useful for problem blocks.

The side-oiler system is substantially different than the common top-oiler FE systems found on the 390, 428, and other FE engines. The side-oiler directs oil to the main bearings first before circulating the oil to the top end of the engine. The point was to keep the main bearings well lubricated during racing conditions. Side-oiler cam bearings have extra feed holes for the heads, which line up with grooves in the cam journal.

Install the engine's "real" cam using oil on the journals. Installing it now allows you to be gentle and smooth on the bearings, minimizing the inevitable bumps from each lobe as it goes into place. It also allows you to install the cam without lube on the lobes, making the job less messy. A touch of lube on the cam front face and you can install the thrust plate. Mount a magnetic dial indicator and check cam-thrust clearance, a quick task with no other parts in the way. The cam should turn easily by hand using the timing sprocket as a handle. Put the proper lube on the cam lobes and go on to the next step in assembly: the crankshaft.

Main Bearings, Rear Seal and Crankshaft Installation

Check bearing clearances using a dial bore gauge. Plastigage is useful, but is not precise or reliable enough for performance work. Install your desired main bearings into the block without a rear seal. Install and torque the main fasteners according to the specification using the desired lubricant. Install the cross bolts in 427 engines, and yes, they indeed do make a difference.

Measure the chosen crankshaft journal with a micrometer. Lock the micrometer in the measured position

FE cam bearings have a common inside diameter, but get smaller in outside diameter from front to rear.

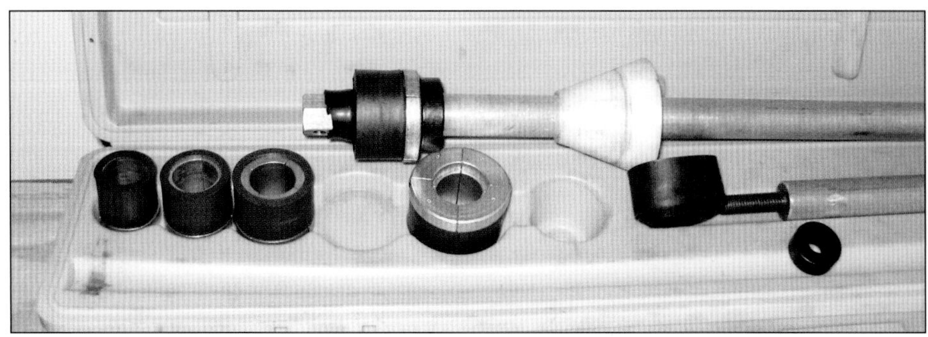

Using a basic cam-bearing installation tool is the most common way to install the bearings.

and lightly clamp it into a vise (aluminum jaws) with the open end facing upward. The dial bore gauge fits into the mounted micrometer and adjusted to read "zero" as you sweep it though the mic's measuring faces.

Now take the dial bore gauge and slip it into the corresponding main bearing bore. It reads clearance directly as the difference between the bearing inner diameter and the crank journal. The target range for main bearings in iron-block engines is .0028 inch with anything between .0025 and .0030 inch considered acceptable. Aluminum blocks are targeted to about .0020 inch because they enlarge as they warm up.

If you come up with improper clearance, disassemble and inspect the suspect bearing and cap before rushing to judgment. The smallest particle of debris caught under the cap's bolt face or the bearing shell impacts the reading.

Clearance adjustment can be handled by touch-grinding the crankshaft, or by using selective-size bearings. Either option is perfectly acceptable and yields good results. I prefer to use selective-size bearings because altering the crankshaft makes future service problematic. You can use mix-and-match selective-size bearings either as a complete set or as half-shells to get the desired clearance.

No manufacturer supplies a .001-inch oversize or undersize bearing for the FE. Fortunately, you can use selective-size bearings for the 351 Cleveland engines with minor work. The only difference is the small locating tang, which is in the wrong position for an FE. Either remove the tang (it's an assembly aid only and does not prevent the bearing from spinning), or add an extra tang slot to the block and/or main cap with a small file or cutoff wheel.

With vertical bearing clearances measured and addressed you can move to the thrust measurement. Remove all the main caps, lubricate the bearings, and install the block half of the rear main seal. Use a really, really thin film of Motorcraft TA-31 silicone on the seal where it contacts the block, and offset the seal edges relative to the cap face by .100 inch. Lay the crank into place and install main caps, except for the rear one. Install the cross bolts for those spots if equipped. You may need to use a flashlight and a plastic mallet to properly position the caps. The cross bolts must go in smoothly. You do not want to draw the fasteners into place; they should thread in with your fingertips. Torque the vertical fasteners on numbers-1, -2, and -4 main to specification, leaving the thrust cap fasteners and cross bolts finger tight.

Set up a magnetic dial indicator and measure thrust clearance. I use a big screwdriver to wedge the crank backward and forward in the block several times with the last motion being forward. If clearance is sufficient (between .008 and .0012 inch), I torque the vertical fasteners on the thrust cap and recheck before moving ahead. If you do not have enough clearance, you need to remove the crank and make some changes.

Motorcraft TA-31 silicone seems to be very durable and oil resistant. It was originally intended for diesel applications.

Some Genesis blocks have the machined chamfer, where the thrust surface meets the crank journal area, cut smaller than stock. This interferes with certain bearings, causing inadequate thrust clearance, as the bearing gets spread out and deformed upon installation. If present, the problem can be resolved by enlarging the chamfer with a sharp file. On some blocks, the outside diameter or the machined thrust area is too small. If interference is found during assembly, it's usually best to reduce the bearing thrust diameter on a lathe, although block modification is possible. If you need to thin the thrust bearing a touch, it can be sanded with 600-grit and WD-40, on a sheet of glass or a granite table. Try to remove material only from the non-loaded front face.

Before installing the rear cap, you need to check a couple things. When using main studs, be certain that they do not protrude beyond the oil pan gasket surface. Otherwise, the studs will interfere with the windage tray or even the oil pan itself. Also do a quick inspection to be sure the drain holes seal into the rear main cap, and verify that they are not blocked off by the pan gasket or windage tray.

For rear main cap assembly I use Motorcraft TA-31 gray silicone. A dab is applied in each corner and a very thin layer on the flat areas between the seal groove and up to the pan rail. Avoid getting silicone near the bearing end of the cap. The FE uses vertical side seals that go into grooves in the main cap. The nails

Camshaft installation is easiest before the crank goes in. You can use both hands to smoothly guide it into place. Be careful to avoid banging it against the bearings.

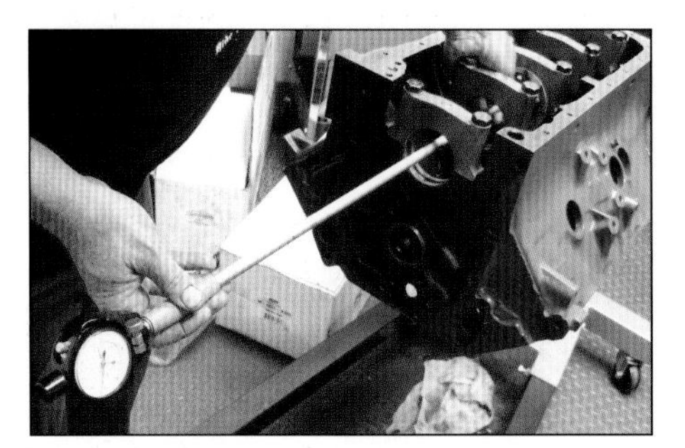

A dial bore gauge is the most accurate tool for checking bearing clearances. And this is an important step in the assembly process. Plastigauge is better than nothing, but is nowhere near as accurate. Once set up, the dial bore gauge method is very easy, quick, and repeatable.

are tapped between the cap and each side seal to force those seals against the block. I use the same RTV silicone as assembly lube for both the side seals and nails.

Install the bearing and the rear main seal. Slip the lubed side seals a little ways into their grooves and wiggle the whole assembly into position by hand. You should be able to start it far enough for the fasteners to be used as guides. Use a plastic mallet to tap the cap into place, alternating with light taps to the side seals so they slide down toward the block. Lube the nails and use a small hammer to tap them down flush with the pan rail. Torque the vertical main fasteners to spec. Assembled properly, the nails are below the pan rail, and side seals are either a tiny amount above or below. A razor blade or a dab of silicone levels the pan gasket surface. You should see a tiny amount of silicone around the cap to block interface on the rear of the engine, which is a visible indication of a complete seal.

I check bearing clearances with a dial bore gauge (this is a working-end view). The gauge leaves a fine line in the bearing that does no harm, but it can be polished out with WD-40 and a bit of brown paper bag material if desired.

The correct rear-side seal orientation is to have the nails against the cap. Starting the seals in first before putting the cap into the block makes things a bit easier.

For the cross bolts, I use the common ARP lube on the threads, but a small amount of RTV on the washers and bolt undersides. This prevents oil seepage. I torque these to 45 pounds working side to side from the center cap out.

At this point, you should be able to turn the crank smoothly by hand with no stiff spots or dragging. If you need to use a wrench on it to get the crank to spin, something is wrong.

Pistons, Rings and Rods Assembly

I suggest you assemble all the pistons, rings and rods as a group before installing any of them into the engine. This work is comparable to that of building any other engine. Nothing is particularly unusual or FE specific about the procedures. I just recommend following good shop practices. The assumption is that all the parts have been cleaned and inspected beforehand. FE engines are pretty forgiving as far as piston-to-valve and piston-to-cylinder-head clearances. But these clearances need to be checked, and you may well find yourself repeating the piston-installation process.

Most if not all connecting-rod manufacturers recommend tightening the rod bolts to torque or stretch specification and loosening them several times before final assembly. This cycling process serves to burnish the threads and respective contact surfaces. Bolts that are cycled several times are more consistent in the way they take up tension and load. Connecting-rod manufacturers recommend one of three ways to determine proper rod-bolt installation: torque, torque angle, or stretch.

A piston-ring filer makes the job a lot quicker, but you can make do with a sharp file clamped in a vise. For somebody doing this for the first time, it's best to trim one end and use the other end as reference to keep things straight.

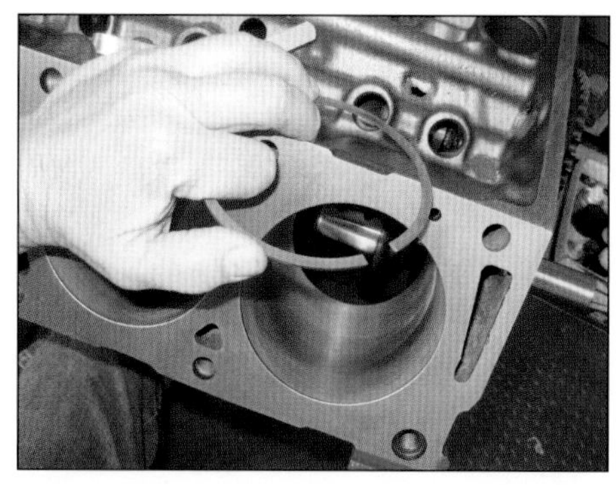

When it comes to ring filing, you want to just take the sharp edges off after sizing the gap. Note the barely visible edge break. There's no need for a big chamfer—it is actually undesirable to have one.

Torquing to a "foot-pounds" specification is the least accurate method of installation, but literally millions of engines have been successfully built with nothing beyond a basic torque wrench. Since torque wrenches measure friction, a defined lubricant and the aforementioned cycling of the fasteners are keys to a successful outcome.

Torque angle is a means to reach a stretch specification using fairly simple tools. It can be very useful in tight clearance areas and where a stretch gauge cannot be used. The fastener is initially tightened to a modest torque value, and then further tightened an additional number of degrees. An example would be a specification of 35 ft-lbs and 45 degrees.

The first step in actual assembly is to measure bearing clearances. Similar to the process used for the mains, the crank is measured with a micrometer, which in turn is used to "zero" set a dial bore gauge. Assemble each rod bearing into a connecting rod, torque the rod cap into place, and use the dial bore gauge to get the reading. You can target a similar .0025 to .0030-inch range for desired clearance. On factory FE rod journals, there are no available selective bearings, which means that machining the crank or rods is your only choice if adjustment is needed. Most stroker combinations use 2.200-inch big-block Chevy-size journals and numerous clearance-adjusting options are available. I also test fit the piston pin in the rod. Clearances here are among the tightest in the engine at .0008 to .0010 inch, and should have already been set at this point by your machine shop. The fit should be glass smooth with no tightness.

Piston preparation involves careful inspection and some detailing. The oil return holes that are in the root of the oil ring groove are a common place to find burrs and chips, so you need to remove these imperfections. Also spend a little time to smooth off any sharp edges around valve reliefs and at the lower edge of the piston skirt. Clean and lubricate the piston pin and test fit it in the pin hole. The pin should slide in and rotate freely with no drag. Piston skirt clearances should have been set by your machine shop, but can be verified by using the same micrometer and dial bore gauge technique used for bearing clearances. Be aware that the use of a torque plate during machining changes the actual measured clearance, but you should be reasonably close.

Installing spiral locks can be a challenge. Pull them apart like a spring and wind them into place, being certain that they are firmly engaged into the receiving groove. Make certain the piston and rod are correctly oriented. You want to get the installation right the first time because these are even tougher to take apart.

Most pistons have a single- or double-spiral-lock retainer. I find it easiest to install the locks on one side first before installing the pin. Installing the spiral locks is a matter of spreading them out a bit and

A tapered piston and ring installation tool is a very good investment. It saves a lot of effort and helps prevent broken rings. Once you try one, you'll never want to use anything else.

winding them into the groove. On an FE you install the pistons with the valve relief pockets on the intake manifold side of each bank. Connecting rods have a large chamfer on one side of the big-end bore, and a small chamfer on the other. The larger chamfer always faces the crankshaft counterweight, and the smaller one faces the adjoining rod.

Hold a piston in one hand, slide the pin partway in, and then put the rod into position. Slide the pin the rest of they way in and install the other spiral lock. The piston rings are installed next, starting with the oil expander, then the oil-ring rails. The second ring is installed next using a ring expander tool, and then the top ring is installed. Do not wind the two upper rings into place. Winding or overexpansion causes the rings to deform and results in a compromised ring seal and reduced ring life. There are a few theories regarding gap placement, but the only really important thing is not to line them up; the rings do slowly rotate when the engine is running.

Although many people and companies recommend a variety of supposedly magical elixirs and powders for prelubrication and break in, I prefer plain old non-synthetic engine oil. Oil is the lubricant from the second the engine fires up until it is removed from service. Any fancy stuff is either washed away during pre-lubrication or instantly dissipates on fire-up. Cylinders get a small amount of oil massaged into the surface. Bearings get a few drops wiped onto them along with a film on the crank journal. I put a bit on the piston skirts and a thin film on the rings. I do not assemble anything dry, but you do not want to dunk the piston or have it dripping wet.

Use a tapered ring compressor. If you don't have one, buy one. Older clamp-style compressors certainly work, but once you've installed pistons with the tapered design, you'll throw your old one away without a moment's regret. Install one piston/rod assembly at a time. Tighten the fasteners to specification and rotate the engine to check for smoothness. If you install a piston one cylinder at a time, it helps to isolate any problems that may occur. As the piston pairs are put together on each journal you can check the rod side clearance. This clearance should be between .016 and .020 inch on the average FE build, but is more forgiving than other clearances in the engine (it runs fine at .012 inch and at .025 inch).

Bring each piston to the top and check the deck clearance. You are again checking the work of your machinist. My preference is to be either "zero" or down from the deck by no more than .005 inch, but anywhere between .003 inch positive and .010 inch below deck runs just fine. The pistons should all be very close to each other in deck clearance; differences of only a few thousandths are really inconsequential. Leave the engine with number-1 piston at top dead center.

Timing Chain, Cam Degreeing and Front Cover Installation

With the cam and crankshaft in the block, next go to the timing set. On the majority of FE projects the basic Cloyes set is pretty intuitive. (I already covered the component selection in Chapter 6.)

The crankshaft sprocket is a light push/press fit on the shaft. Cam and crank sprockets go "dot to dot" for setup. With both cam and crank at

Take a look at one of our Engine Masters Challenge entry's bottom-end components. It gives you an idea of the level of modification that can be done to a factory steel crank in the pursuit of horsepower.

A large degree wheel is a very useful tool when setting cam timing because it is easier to see and make small changes.

Using a damper spacer as an alignment sleeve for timing cover positioning is usually a good technique for most builds.

their "number-1, top dead center" positions and the crank key installed, put the crank sprocket on and slide it back within 1 inch of its final position. Take the cam sprocket, loop the chain over it, and hook them over the crank sprocket. You should be able to use the cam dowel pin as a guide to push and tap the whole assembly into final position, working alternately from top to bottom. A plastic mallet and a piece of aluminum tubing over the crankshaft snout come in handy. This should only take a light touch, not a heavy beating. Temporarily install the cam bolt and washer.

At this stage, you need to degree the cam. Install the degree wheel on the snout of the crank, and make a pointer from a piece of welding rod to indicate TDC. Install the appropriate lifter in the number-1 intake position. I use a magnetic-mounted dial indicator to read the lifter movement. Since cam lobes may be asymmetrical, you cannot simply assume that the point of highest lift is the centerline. To determine the intake lobe's centerline, you need to find the spot .050 inch below maximum lift on either side of the lobe's peak and then split the difference between them on the degree wheel.

Experience tells you whether the engine wants a particular cam installed at any given degree position. Lacking direct experience, it is usually best to use the suggestions from the cam grinder as the first point of reference. If you are satisfied with the installed intake centerline, you can now continue to rotate the engine and check intake and exhaust opening events against the cam-card specifications. The Cloyes timing sets allow changes in increments of 4 degrees. Based on dyno testing, changes of only 1 or 2 degrees seem to have a very nominal impact on performance and are not worthy of much concern on the first try of a cam combination. If everything checks out, remove the degreeing hardware.

Next, install the cam bolt, washer, and fuel-pump eccentric. Use red Loctite and torque to specification for a secure fit.

The front seal should be installed into the cover and lubricated. Glue the timing-cover gasket into place. Slide the sheetmetal oil slinger onto the crank snout. Place the cover in position using the crank damper spacer as a centering guide. Most of the timing-cover bolts go into either water or oil, so Teflon paste should be applied to the threads before installation to inhibit corrosion and ensure an accurate seal is attained.

With the cover bolted on and the damper spacer slid into place, you can install the damper itself. Check the fit by measuring its inside diameter and comparing that to the crank snout. It needs to be a press fit, but not be too tight. Each damper manufacturer has a desired specification and honing may be needed. The damper key is a simple 1/4-inch-square stock, and needs to go in after the damper spacer is installed. You should be able to start the damper installation with a large dead-blow plastic mallet, but you need to use a press or bolt to complete the installation. A threaded stud and a bearing washer/nut is the right way to do this. Because the FE damper bolt is both large in diameter and long, plenty of them have been installed

Valvespring pressure testers are a necessity in a professional shop, but they are very expensive. Checking and adjusting spring pressure is a task that your shop should handle as part of the cylinder head assembly work.

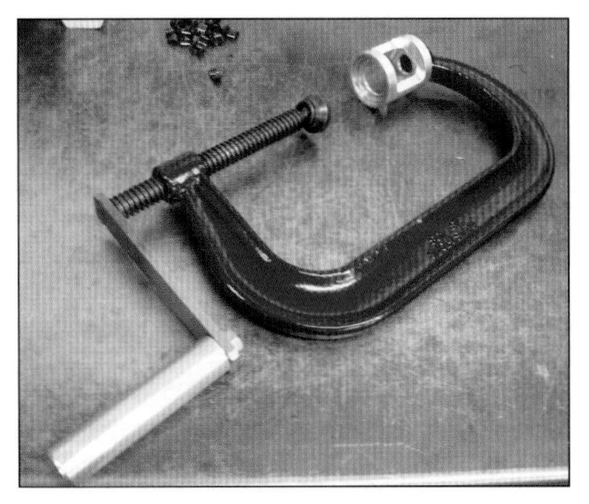

The valvespring installation removal tool from Mondello Performance Products is almost embarrassingly simple and effective at removing almost any valvespring.

using the bolt itself to pull the damper into place. I cannot remember ever seeing a stripped or broken FE damper bolt.

You do need to use the proper thick washer and to torque that big bolt to specs, or it will come loose.

Cylinder Head Assembly

Often cylinder heads are assembled before the block is completed. Springs and retainers are chosen to meet the needs of the camshaft and application. I strive to get installed heights within about .050 inch of coil bind. On solid-roller cam applications, a stiffer spring causes fewer problems than one that is too weak, especially in regards to seat pressure.

On an FE engine it is important to get the valve-tip heights close to equal relative to both the head's deck and the rocker mounts. Significant differences in tip height require uneven pushrod lengths and cause potentially odd geometry issues. In addition they require a wide array of shims to equalize installed heights for the springs. They run okay in the farm truck, but are decidedly not desirable in any kind of performance use.

You need to prevent any potential piston-to-valve contact. While test assembling the top end is time-consuming and tedious, it is very necessary to verify operating clearances because bent valves and damaged pistons are very expensive to replace.

Factory FE heads used umbrella-type valve seals, but everything today uses a positive viton seal that presses onto the guide boss. You need to establish installed height and position the requisite shims and the spring locator before installing the seal. Install the valves using oil as lubricant. In a commercial shop, an air-powered spring compressor is used, but the Mondello "big purple C-clamp" is startlingly effective for low volume and home use. Available through Goodson or Mondello, the clamp is simple, portable, and safe—much safer than the old-style, over-center hand-lever compressors.

Cylinder Head Installation

Clean the decks of the block with thinner or brake cleaner. Install the head-locating dowels by tapping them into place with a small hammer. Starting with the split sides in first on a slight angle make this a bit easier. If using studs, install them now with a touch of ARP lube, lightly seating them in the block.

On engines utilizing domed pistons, paint a couple piston tops with magic marker or machinist's dye and mount the heads without gaskets. Turn the engine over through a couple cycles. It should rotate smoothly with no interference. If it clicks or binds, the interference points are clearly visible as witness scrapes in the marker or dye. Modify the heads and pistons as needed and try again. Smooth, interference-free rotation means you have at least .041 inch (a gasket's thickness) of clearance.

Next, check for adequate piston-to-valve clearance. Install head gaskets with the coolant openings facing forward on both decks. There is no correct "up" or "down," but there is a definite "front" and "rear." I prefer using clay to do the check because it allows me to see both vertical and radial clearance in one test. I first use silicone-spray lubricant on the piston dome and the combustion chamber to prevent the clay from sticking. Then I pack both valve pockets with clay and install the head using a couple head fasteners. Lifters, some reasonably close pushrods, and the rockers for the test cylinder follow. Fastener torque and valve adjustment need not be perfect, but should be close for accurate results.

Rotate the engine through two complete revolutions. The clay puts

FE head dowels should be replaced on every build. They are quickly installed by tapping them into place by starting the split side in on an angle.

The preferred head gasket is the Fel-Pro 1020. The water opening always goes to the rear, and the compression reinforcement appears to be "up" on one head and "down" on the other. Each gasket is clearly marked "FRONT."

The Fel-Pro gasket must properly line up with the oil passages on the Genesis block, otherwise no oil will route to the rocker arms. And once again, if this critical procedure is not done properly, the engine can experience serious damage.

up a little resistance but you should not have to force it. If you hit a hard stop, take things apart to find out why. Remove all the valvetrain parts and the cylinder head. The clay should show a clear impression of

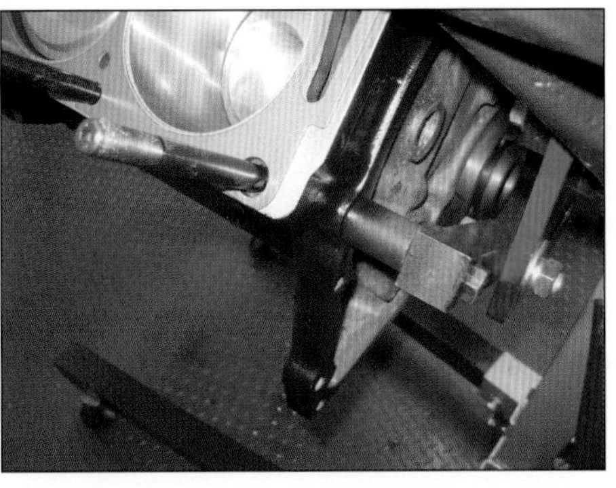

The rear corner of the head gasket is shown correctly installed.

This front corner of the head gasket is also correctly installed.

each valve. I use an X-acto knife to take clean-edged pie-shaped cuts from the clay. This allows me to see clearance both vertical and radial around each valve. The standard desired measurements of .050-inch radial and around .100-inch vertical apply. With the exception of maximum-effort race applications, FE engines rarely have problems vertically, but radial clearance can be an issue with some pistons. Obviously if you do not have enough, the pistons are coming back out for machining.

If all the clearances are okay and everything looks good, the heads can be bolted down for good. If you have a Genesis block, be sure that the oil-feed hole to the heads lines up with the head-gasket opening. It is common to need an extra oil hole drilling in the gasket. Be sure to use lube on all threads, as well as on the

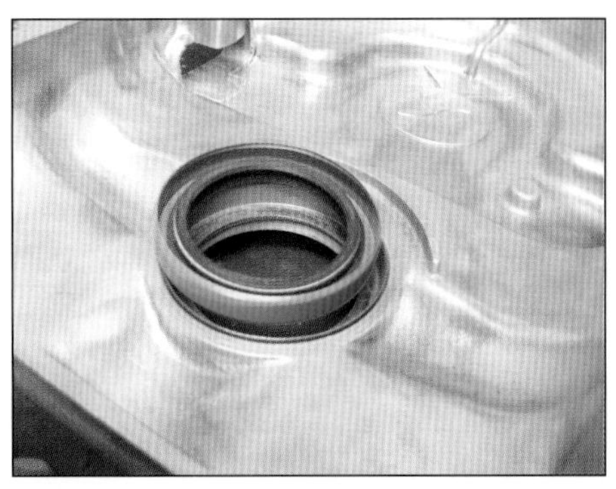

Here is the correct front seal orientation. The seal must be installed before the timing cover goes on.

underside of the head-bolt washers and the nuts or bolt heads. Follow the factory torque pattern, which is essentially a spiral from the middle outward. Torque should be the same as was used during torque plate honing. Go to a light torque value to seat the head, then run the pattern going to full torque on each fastener in a single full sweep. Walk away from the engine for an hour and then do a cold re-torque. This allows the head gasket and fasteners to relax and take an initial set. Loosen each fastener and retorque it again in a full sweep. If you paint mark the fastener you'll find that this process gets another few degrees of rotation for a given torque value.

Lifters and Pushrod Length Check

Install the lifters after using the manufacturer's specified cleaning and lubrication process. On flat tappets, this means using a high-pressure lube on the contact face and oil on the sides. With rollers, each supplier has a different procedure. Some want a full wash followed by an oil bath, while others use a particular assembly grease that they do not want washed away.

The MS90145 Fel-Pro low-riser passenger-car gasket is a durable part, but I usually use a thin amount of sealer as well to form a very solid seal.

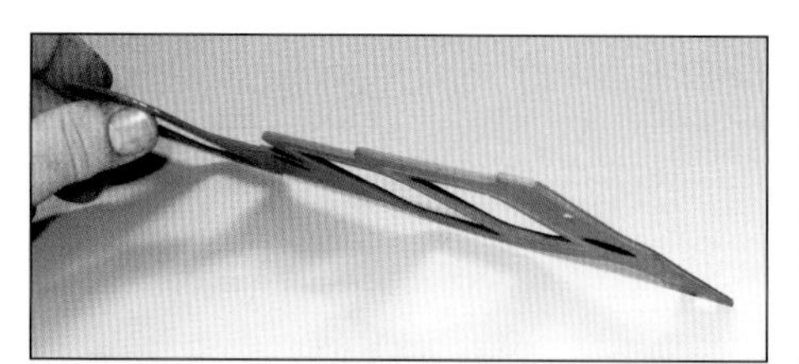

A new intake gasket with a steel liner is far more durable in street use and is recommended for non-ported Edelbrock or medium-riser heads. Survival keeps these problem-solver gaskets in stock.

A Fel-Pro 1247 medium-riser race gasket is easy to trim for altered ports, but it won't hold up in street operation. Oil saturation causes it to deform and oil gets pulled into the ports.

With lifters installed, set the cam to the base circle, which is the lowest lift point, and again install the rocker assembly on one head. Adjust the adjuster so that only two or three threads show below the rocker arm body. Use a traditional ball/ball-style pushrod-length checker and extend it, taking up all of the lash in the valvetrain by touching against the ball end of the adjuster. Unbolt the rocker assembly, remove the checking pushrod and measure it using a 12-inch dial caliper. When you call your chosen pushrod supplier and specify the length as "bottom of cup," they will know exactly what you need. Traditional FE engines use a 3/8-inch cup at the rocker end, which is different than other engines. Be sure to specify that when ordering.

Every engine combination is a little different, but common pushrod lengths run between 8.900 and 9.100 inches on roller cams, and between 9.100 and 9.350 inches on flat tappets. If you come up with something radically different, you'd best recheck the measurements.

Intake Installation

In Chapter 10, I described the detailed process to verify and correct intake manifold fit on an FE. Now you finally get to install it "for keeps." Use a very thin layer (not even a bead; more like a translucent coat of paint) of TA-31 RTV silicone around all the water openings and the intake ports. Install the gasket, which should have been trimmed to match the ports during the fitting process. Add another really thin layer of silicone on the gasket and a fat bead on the front and rear seal areas on the block. Carefully set the intake into place and use the distributor as a locating dowel. Use a plastic mallet to move the manifold around as necessary for fastener and port alignment.

Install the fasteners being certain to get them all started by hand. If you are using a fuel log on a 2x4 setup, don't forget to install it now. There is a factory torque value for intake bolts, but you almost never get to use it. Truth is, many of the

Trimming the intake gasket at the valve-cover surface is common practice in order to achieve a good cover seal. Use a touch of silicone to prevent any leakage.

Installing the distributor in its proper location is a key to getting proper intake fit.

When the distributor has been correctly installed, you can use its location as a reference point to ensure that other intake measurements are much more accurate.

Note the intake pushrod hole clearance. You want to rotate the engine through several cycles during test assembly to verify smooth operation of every pushrod at all valve-opening positions. Hard contact when the engine is running causes the pushrod to pop off the rocker arm.

I often need to modify the pushrod holes for clearance. Typically, they get checked, marked where they hit, and rechecked after altering the holes on a Bridgeport mill. You can use a die grinder to do the same job by hand.

fasteners on aftermarket intakes are very hard to reach even with a box-end wrench. Working from the inside out to the ends, snugly tightening the fasteners by hand is fine. If you use a flashlight to sight down alongside the ports, you'll see the tiniest amount of silicone squeezed out, which is your visual indication of a good seal. Remove the distributor and be sure that the silicone in the front wall has not pushed into the opening. After several hours you

This shows the distributor hole and the oil galley plug that is overlooked most often. Leaving this plug out results in nearly zero oil pressure during prelubrication. Not getting it tight enough causes interference with the distributor, either preventing installation or binding it up to make timing changes impossible.

This is a manifold marked for milling to correct angle and height variances. If everything is measured properly, a single milling operation can correct numerous dimensional variances.

can use a razor blade to trim the silicone on the front and rear seal area for a more professional appearance.

Valvetrain Assembly

The rocker arm assembly on an FE is unique, even in installation. I really prefer using studs to mount the rocker assemblies. They are easier on the threads in the cylinder head and make installation simpler. If you choose to use bolts, be certain that you have the correct length in each position. If they are too long they bottom out giving you a false torque reading, and may pull the threads out of the head when running. On Edelbrock heads, if you use the wrong-length stock bolts, they break through the top of the port.

The first assembly step is to lube the ends of the pushrods. Slip each pushrod through the hole in the intake and center it in the lifter. Using a flashlight helps to be sure they end up in the right place; it's easy to miss. Set all the rocker adjusters so that they are showing two or three threads below the rocker body. If you are using the factory tin drain trays, set them in place. Now set the rocker assembly into place by lining up the pushrod cups with the arm adjusters. It is a bit of a juggling exercise to keep all the pushrods in place while starting the fasteners. Tighten the fasteners a small amount at a time moving from one to the other in order to prevent bending the shaft. Go slowly and continue to monitor the pushrod position as you tighten them. Once they are all snug and the rocker stands are visibly in contact with the head, you can bring them up to torque.

When both rocker assemblies have been installed, you can rotate

Note the pushrod-to-intake clearance. Once the intake is bolted down, this cannot be easily altered if it the clearance is inadequate. Miss it and you'll be taking it back apart again.

Note the orientation of the fuel log and oil-return tin. The log must be mounted before tightening the intake, and the tins don't work on all rocker or manifold combinations.

Check out the carb-spacer-to-gasket overlap. You can spend a minute of time to trim for better fit and you can find a bit of free power in the process.

the engine. Look, listen, and feel for any interference. The engine is a bit harder to turn but should still be smooth and quiet. Clicks, snaps, and tight spots need to be investigated and corrected.

Valve lash for solid lifters (or preload for hydraulics) can now be set. Most solid lifters specify a hot lash setting. Since the engine is cold you need to do an initial approximate setting. On iron block-and-head combinations, there is roughly .004-inch growth in lash from cold to hot. On engines using iron blocks and aluminum heads the growth is around .009 inch. An aluminum block and heads gains somewhere around .016 inch in lash as they warm up, which often requires nearly zero cold lash.

The same growth factor applies to hydraulic-lifter preload. Therefore, if the cam manufacturer wants .020-inch hot preload, you need to provide a greater amount when cold. With some performance hydraulic systems running near zero lash, inadequate cold preload results in a noisy valvetrain when warmed up.

Oil Pump, Windage Tray and Oil Pan

When installing the oiling hardware, the first step is to test fit the pump driveshaft into the distributor. Ive seen cases where they do not fit due to burrs or damage. It is far easier to check and correct this before the pan is on.

If you are using a windage tray, test fit it next. Glue a pan gasket to it and bolt it onto the block with just a couple fasteners. Rotate the crank to check for clearance. Try the dipstick. Sometimes they interfere and some trimming may be required. If all is good, remove the tray and put a dab of silicone in the spots where the timing cover and rear cap meet the block. All that's required is a very thin smear along the sides, then remount it again using only a few bolts on each side.

Next, slip the driveshaft into the oil pump with the retainer clip toward the upper end. The pump gasket goes on and the two 3/8-inch bolts get a dab of red Loctite before snugging the pump up onto the block. While most builds will use common automotive bolts, we have used studs and safety-wired nuts on road-race applications. We use an extremely small layer of silicone on the pump inlet gasket to prevent any possibility of air intrusion. The pump pickup is mounted using two 5/16-inch fasteners that have either Loctite or safety wire as insurance against loosening. Some pickups may require modification to clear the windage tray—a test fit is a good idea before final installation.

Installing the rear breather onto the intake is sometimes saved for last because you can easily remove it for quick inspections for a water leak or swap out a noisy lifter.

Now check pan-to-pickup clearance. Instead of clay I use a straightedge (or ruler) and a 12-inch dial

You need to use the correct screw-in core plugs for the engine or the engine could suffer a catastrophic failure. The correct Ford or reproduction plugs provide a proper seal and adequate clearance. However, hardware store screw-in plugs can contact the cylinder before they are tight enough to seal.

A quick screw-in core-plug tool can be made from a deep socket and a nut. With a tool such as this, you can ensure the core plug is properly installed and be sure to use Teflon paste.

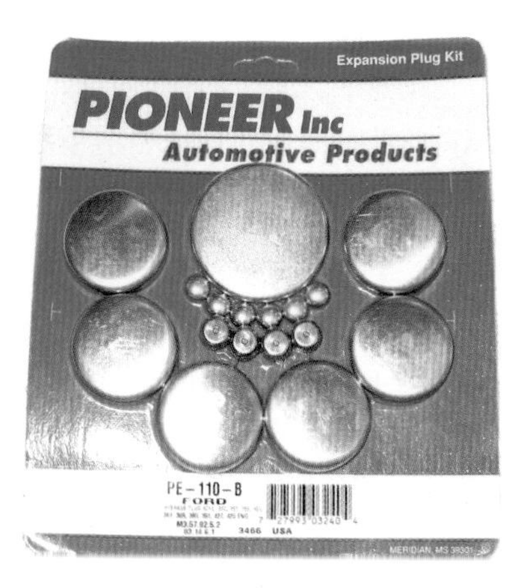

When using common FE core plug kits, check the sizes—1.750-inch diameter is too small for many blocks. The correct diameter is $1\frac{49}{64}$ inches.

caliper. Clamp the straightedge to the bottom of the pickup screen. Measure the distance from the straightedge to the windage tray's pan-rail surface. Lay the straight edge across the gasket rail on the oil pan with a gasket in place. Use the dial caliper to measure from the edge to the bottom of the pan's sump. Subtract the first measurement from the second one and you have the clearance. Somewhere between 3/8 and 1/2 inch is usually about right.

Wash the pan thoroughly before installing it. This is a great time to fill it with mineral spirits and check for any leaks. If all looks good, you're ready to mount it to the block. Put a gasket on and hold the pan in place just below the windage tray. Remove the fasteners holding the tray on (it helps to have an assistant) and put the pan up to the block capturing the tray. The sealer you applied earlier keeps that tray from coming loose during the process. There are 20 bolts holding the pan on, and you want to get them all started before tightening any of them. Once all bolts are in, work your way around

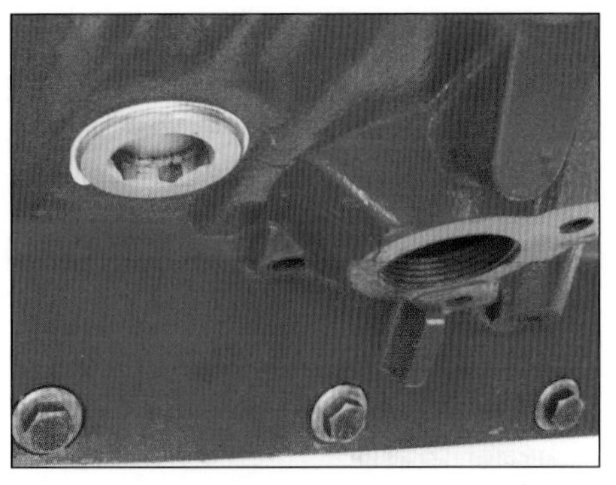

Once the screw-in core plug is in place, it should be installed a touch deeper than this example. If the plugs stick out too far from the block, motor-mount interference can result.

the pan, snugging them a bit at a time. With the sandwich of two gaskets and a tray you do not want to overtighten and risk deforming the pan rail.

Bolt the oil filter mount to the side of the block, add a filter pre-filled with oil, and install an inexpensive pressure gauge into the 1/4-inch NPT tapped pressure opening on the top of the mount. Pour 4 quarts of break-in oil into the pan through the distributor opening. Using a 1/4-inch-deep socket and a long 1/4-inch-drive extension in a drill, spin the pump counterclockwise. You should see around 75 pounds of pressure on the gauge in a minute or so. Oil should come up through the rockers and shafts shortly afterward. When you see oil from all sixteen rockers, you're all set.

If no pressure happens, it's usually due to a missing oil-galley plug. Spin the pump and you see a missing one behind the distributor as a stream of oil though the opening in the intake. Missing the one behind the timing gear is seen as a waterfall viewed through the fuel-pump opening on the side of the timing cover. I know this because I've done both.

Assuming that everything went well, you can install the valve covers. Knowing that you'll be doing an inspection and hot lash, no sealant is used at this time.

Carburetor, Water Pump, Distributor, Wires and Plugs

You are finally on the home-stretch, installing the last group of external parts.

The distributor is dropped into place after locating the number-1 cylinder on the compression stroke. Lubricate the drive gear and the seal. I usually start by setting it at 20 degrees before TDC as a fire-up position. Once close to position, you will likely need to rotate the engine around a bit to get it to drop all the way down onto the pump shaft. The body where the clamp goes should be nearly flat against the intake. If you can see the side of the seal, it's likely the distributor is still not seated.

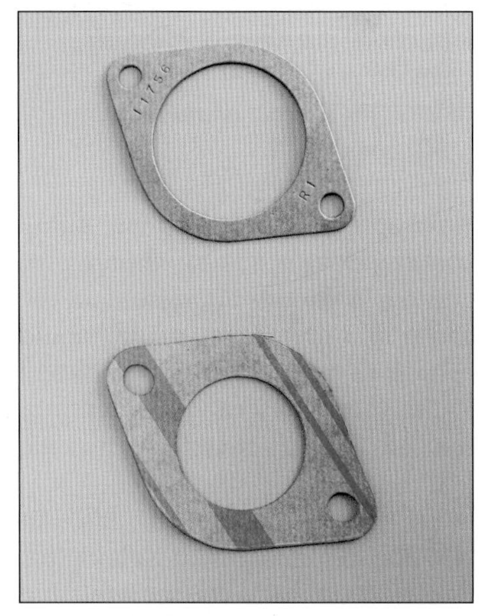

There are different water openings and thermostat gaskets. The early ones are large, and the later ones are smaller. Remember that your water outlet, thermostat, and manifold must match.

Plugs and wires are pretty basic. There are a few tips worth mentioning: Use a small amount of anti-seize lubricant on plug threads going into aluminum heads. Also use a bit of dielectric grease on the plug wires, on the plug boots and at the distributor cap. Route the wires away from linkage and heat, keep number-7 and -8 separated, and keep them away from the distributor magnetic-trigger leads.

Carbs are a simple bolt-down with a couple caveats. Check for clearance throughout the throttle travel. Some carbs interfere with the intake at the throttle plates. Others hit the throttle linkage. Still others might have a vacuum passage that overhangs the intake's mounting flange and leak. Carburetor spacers usually solve these problems, but you need to check. The factory throttle linkage on a 2x4 setup is designed to be set for wide-open throttle, not to vary the progressive opening of the two carbs.

The assumption here is that you are going to put your engine on the dyno for testing and tuning. This requires only the basics as far as front-end dress.

Double check to be sure you can see the head-gasket corner tab in the front. That indicates that the gasket is properly installed.

A mechanical water pump is a simple bolt-on item that requires four fasteners with Teflon paste, one bypass hose, and you're done. The damper-mounted pulley, a water-pump pulley, and a short belt complete the test package. Working with an electric pump is even easier because the bypass opening on the intake is plugged and a belt is not needed. For dyno testing, a water outlet is used but no thermostat is installed.

Stand back and admire your work. Now, you're ready to fire this engine up!

Max-Performance FE Engine Builds

As I've detailed, there are many different head, intake, block, and rotating assembly combinations that produce respectable and reliable power. At Survival Motorsports, we've built dozens of FE engine combinations. On page 139 are two engine recipes that provide a sound parts combination for a fast and humane max-performance FE Engine.

Page 139 (top left): This stroker provides exceptional torque and horsepower on the lower end of the powerband. Therefore, it is ideally suited for the street. The engine delivers 510 ft-lbs of peak torque at 4,200 rpm. As you can see, the engine consistently builds horsepower all the way up to its 490-hp peak at 5,200 rpm.

Page 139 (top right): This particular 482 Roller Cam FE powerband is somewhat similar to the 390 stroker combination. It produces maximum torque of 620 ft-lbs at 4,300 rpm and it hits a maximum 613 hp at 5,800 rpm.

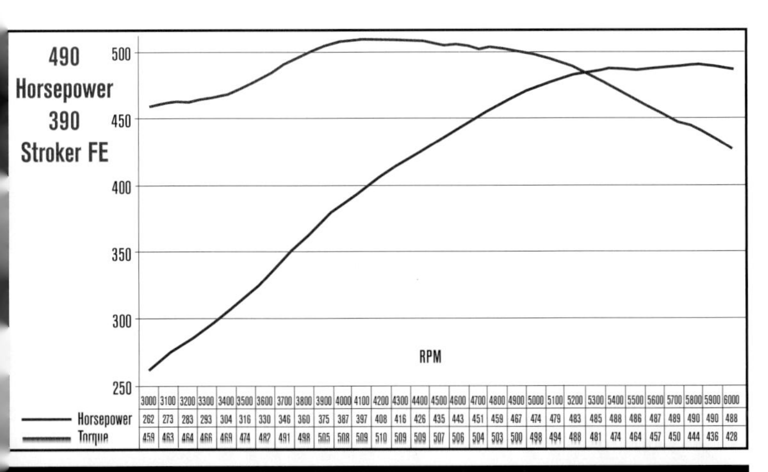

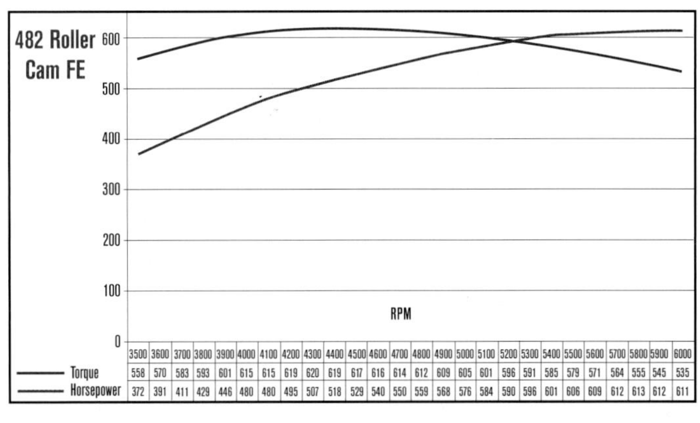

Ford 390 Stroker Build Recipe: 490 hp, 503 torque

Part Number	Description	Quantity	Unit Cost $	Total Cost $
	390 FE Block	1	200.00	200.00
	Edelbrock heads – Bowl porting and prep by Survival Motorsports, 2.200/1.650 stainless valves, 4130 retainers, double springs, keepers, locators, seals	1	2,200.00	2,200.00
	Balanced rotating assembly – Scat crankshaft (4.250 stroke), Scat I beam connecting rods (6.700 long), Diamond pistons (target compression 9.8:1), Total Seal piston rings	1	1,950.00	1,950.00
125M, 8-7200CH	F-M race bearings	?	?	?
7105	Edelbrock Performer RPM 4-bbl intake	1	339.95	339.95
0-81770	Holley 770 Street Avenger vac secondary	1	375.95	375.95
	Carburetor spacer – 1 inch	1	24.95	24.95
CL33-245-4	Comp Cam/Lifters – Solid 236 duration .571 lift	1	258.00	258.00
1445M	Cam bearings – Federal-Mogul	1	25.00	25.00
80009	Damper – non-SFI-Professional Products	1	98.77	98.77
	Damper spacer – billet – Blue Thunder	1	60.00	60.00
155-3601	Fasteners – ARP heads	1	62.99	62.99
6009	Fasteners – ARP rocker studs	1	102.69	102.69
	Fasteners – remainder of engine – OE	1	20.00	20.00
1020	Gaskets – Fel-Pro cylinder head	2	36.95	73.90
1847	Gaskets – Fel-Pro oil pan	1	15.95	15.95
2720	Gaskets – Fel-Pro completion set	1	31.95	31.95
1632	Gaskets – Fel-Pro valve covers	1	19.95	19.95
1247 S3	Gaskets – Fel-Pro intake manifold	1	33.90	33.90
1442	Gaskets – Fel-Pro header/exhaust	1	21.95	21.95
	Duraspark Ford distributor	1	130.00	130.00
FD-129	Distributor cap – Standard	1	24.95	24.95
FD-311	Distributor rotor – Standard	1	5.99	5.99
	Distributor hold-down – OE	1	10.00	10.00
35383	MSD Plug Wires – Black	1	95.95	95.95
	Oil – Brad Penn break-in 30W	8	2.99	23.92
FL-1A	Oil Filter – Motorcraft	1	5.99	5.99
	Oil filler caps – OE replacement	2	6.95	13.90
30740	Oil Pan – Milodon	1	129.95	129.95
M57HV	Oil Pump – Melling	1	46.95	46.95
18640	Oil Pump Pickup – Milodon	1	39.95	39.95
154-7902	Oil Pump Driveshaft – ARP	1	18.95	18.95
	Paint – Duplicolor Ford Blue	1	5.95	5.95
	Pushrods – custom length	1	195.00	195.00
	Roller Rocker Assembly – Erson	1	795.00	795.00
C61YC	Spark Plugs – Champion	8	2.99	23.92
9-3108	Timing Set – Cloyes Tru-Roller	1	120.00	120.00
	Timing Cover – OE	1	35.00	35.00
	Valve Covers – OE	1	35.00	35.00
	Total Parts			**7,402.27**
Services	**Description**	**Quantity**	**Unit Cost $**	**Total Cost $**
	Engine machining – Bore and hone with torque plates, line hone, parallel deck, Install cam bearings, etc.	1	800.00	800.00
	Assembly – Complete engine	1	1,200.00	1,200.00
	Port Match intake	1	300.00	300.00
	Dyno testing	1	1,000.00	1,000.00
Notes:	**Total Services**			**3,300.00**
	Total Parts and Services			**10,702.27**

Ford 482 Stroker Build Recipe: 613 hp, 503 torque

Part Number	Description	Quantity	Unit Cost $	Total Cost $
	Genesis Iron 427 FE Block w/Billet Main Caps	1	3,900.00	3,900.00
	CNC Ported Edelbrock heads – Full CNC porting and prep by Survival Motorsports, 2.200/1.710 stainless valves, titanium retainers, Manley Nextek springs, keepers, locators, seals	1	3,669.00	3,669.00
	Balanced rotating assembly – Scat crankshaft (4.250 stroke), Scat H-beam connecting rods (6.700 long), Diamond pistons (target compression 11.2:1), Total Seal piston rings	1	2,300.00	2,300.00
125M, 8-7200CH	F-M race bearings	?	?	?
2937	Edelbrock Victor 4500 4-bbl intake	1	545.00	545.00
0-8082-1	Holley 1050 Dominator carb	1	375.95	375.95
	Carburetor spacer – 1 inch	1	24.95	24.95
	Custom Solid Roller Cam/Lifters – 260/266 Duration .695/.695 Lift	1	757.00	757.00
1445M	Cam bearings – Federal-Mogul	1	25.00	25.00
	Damper – SFI – Romac	1	295.00	295.00
	Damper spacer – Billet – Blue Thunder	1	60.00	60.00
155-4201	Fasteners – ARP head studs	1	134.99	134.99
6009	Fasteners – ARP rocker studs	1	102.69	102.69
555-9602	Fasteners – ARP stainless, remainder of engine	1	119.95	119.95
1020	Gaskets – Fel-Pro cylinder head	2	36.95	73.90
1847	Gaskets – Fel-Pro oil pan	1	15.95	15.95
2720	Gaskets – Fel-Pro completion set	1	31.95	31.95
1632	Gaskets – Fel-Pro valve covers	1	19.95	19.95
1247 S3	Gaskets – Fel-Pro intake manifold	1	33.90	33.90
1442	Gaskets – Fel-Pro header/exhaust	1	21.95	21.95
8594	MSD Billet Distributor	1	295.00	295.00
	Distributor Hold Down – Stainless	1	25.00	25.00
35383	MSD Plug Wires – Black	1	95.95	95.95
	Oil – Brad Penn break-in 30W	8	2.99	23.92
FL-1A	Oil Filter – Motorcraft	1	5.99	5.99
	Oil filler caps – Blue Thunder Comp Style	1	71.00	71.00
20608	Oil Pan – Moroso T type race	1	308.00	308.00
M57HV	Oil Pump – Melling	1	46.95	46.95
26843	Oil Pump Pickup – Moroso	1	49.95	49.95
154-7902	Oil Pump Driveshaft – ARP	1	18.95	18.95
	Paint – Duplicolor Ford Blue	1	5.95	5.95
	Pushrods – C+B6ustom length	1	195.00	195.00
	Roller Rocker Assembly – Erson	1	795.00	795.00
C61YC	Spark Plugs – Champion	8	2.99	23.92
9-3108	Timing Set – Cloyes Tru-Roller	1	120.00	120.00
	Timing Cover – OE	1	35.00	35.00
	Valve Covers – Blue Thunder Comp Style	1	237.00	237.00
	Total Parts			14,859.71
Services	**Description**	**Quantity**	**Unit Cost**	**Total Cost**
	Engine machining – Bore and hone with torque plates, line hone, parallel deck, install cam bearings, etc.	1	800.00	800.00
	Assembly – Complete engine	1	1,200.00	1,200.00
	Port match intake	1	300.00	300.00
	Dyno testing	1	1,000.00	1,000.00
Notes:	**Total Services**			3,300.00
	Total Parts and Services			18,159.71

Engine Break-In and Tuning

I reinstall the inner springs after a 20-minute cam break-in period. Note the clear centered contact pattern on the valve tip, which is indicative of proper valvetrain geometry.

Let's assume that the break-in process and tuning are being done on an engine dynamometer. A dyno test on a new engine build is highly recommended, although not absolutely necessary. Beyond the obvious benefit of identifying the actual powerband of the engine, the dyno allows full access to the engine. This makes spotting any problems, required repairs, and adjustments far easier than if the engine were in the car. Most at-home mechanics or enthusiasts do not have their own dyno or hands-on access to one. However, the tuning and set-up information in this chapter gives you the knowledge to properly guide your engine through the dyno tuning process, so you wind up with the strongest-running engine.

Connecting the Engine to the Dyno

Mounting the engine to the dyno is similar to, but simpler than, installing it into a car. With the DTS dyno, the dyno I'm most familiar with, the engine is mounted to a portable cart and then the engine is connected to the dyno. The engine is attached to the cart with a large plate bolted to the rear bellhousing surface of the block, along with a pair of adjustable legs serving as motor mounts up front. A manual-transmission SFI-legal flywheel is attached to a coupler for connecting to the dyno itself. With the engine physically mounted to the cart, it is wheeled into the dyno room and attached to the machinery.

When connecting the engine to the dyno, you need to hook up water for cooling and install a water-temperature sensor. A modified drain-plug sensor needs to be installed to measure oil temperature. The throttle linkage is attached and checked for WOT actuation. Headers are routed to exhaust plumbing in the dyno room. My dyno headers also use a pair of wide-band oxygen sensors to monitor fuel mixture—one in each collector. Most dyno rooms have an integrated fuel-supply pump and an ignition box, so all you need to connect are the trigger leads for the distributor and the fuel lines onto the carb.

Each time the engine is run through the RPM range on the dyno, it is referred to as a "pull." The dyno controls the amount of load on the engine and the rate of acceleration through a combination of electronics and hydraulic pressure.

Before making any engine pulls on the dyno, you need to perform some final procedures and checks. Hook up a timing light, check for fuel pressure and leaks, ensure that the engine is full of the proper break-in oil, and check for distributor position and spark. In addition, you need to check the oil to be certain that there are no internal water leaks. I also assume that if you are running a flat-tappet cam, the inner valvesprings have been removed for proper break-in—it's an absolute must.

Breaking It In

If everything is set correctly, the engine should fire up instantly after a couple pumps of the throttle. It is particularly important on flat-tappet-cam engines that you do not spend a lot of timing cranking; you want all that cam lube to stay in place for the first few revolutions.

Once it lights up, try to quickly get the engine up to around 2,000 rpm and start checking on the operation of the engine. The base timing needs to be set. Small drips are not an issue, but big leaks or unusual noises are good reason to shut the engine off and fix them. Shutting the engine down does not hurt the break-in process, and is much smarter than continuing to run with a major problem.

Once satisfied that everything looks and sounds healthy, put a light load on the engine and let it run for about 20 minutes. You need to place a load on the engine to help the rings seat in the cylinder. Vary the load and the RPM a bit during this time, and keep an eye on temperatures and oil pressure. Walk into the dyno room while the engine is running and look for leaks, listen for noises, and feel for vibrations.

When the break-in period is over, I usually do a very short dyno test pull from 2,000 to perhaps 4,000 rpm just to get a starting reference for fuel mixture. Remember that you only have the outer valvesprings if you are doing a flat-tappet-cam break-in, you cannot rev it up very high without risking damage.

With the initial run completed, it's time to get the engine ready for the real work. Remove the oil filter and cut it open to check for debris. A few speckles is normal; a damaged cam fills the pleats of the filter with shavings. A magnet can help identify anything you find; bearing and piston material is nonmagnetic. If you find specs of material and minimal debris, you should install a new filter filled with oil and proceed with the dyno run.

The next step is to remove the valve covers, check your lash settings, and then reinstall the inner valvesprings on flat-tappet-cam engines. You need to reset lash again after the engine has been run. At this stage, verify that the engine has been correctly assembled and there are no problems. This is your chance to fix any small leaks you have spotted. Most are quickly cured with a dab of teflon paste or the turn of a wrench.

Tuning for Power and More

An engine dynamometer is great at testing for WOT. It's reasonably good for testing steady-state/steady-load RPM characteristics. And it's darn near useless at testing for transient throttle response and behavior. Although an OEM has expensive and sophisticated software-driven dyno cells to do such testing, you're not going to find one in any average race shop this side of a NASCAR team. You are testing for a baseline, with the expectation of final tuning in the car itself.

Tuning, whether on a dyno or in a car, is a three-step process. First, you need to establish control. In the case of our subject engine this means a safe and repeatable fuel curve as defined by the dyno's oxygen and BSFC data. The second step—with control established—is to look for trends. Trends are defined by making a single change and gauging the results of that change. If the engine responded favorably (increased horsepower or torque), you should go further in that direction and test again. If the engine didn't respond well to the change, go in the opposite direction and test. It takes at least three dyno pulls to define a trend.

Keep in mind: you cannot extrapolate information beyond the range of your tests. If 32 degrees of timing does better than 30 did, it doesn't mean that 34 is better yet. The last step is to let the engine decide what it "wants." If best power calls for goofy timing or fuel numbers, the odds are that something is wrong with a sensor or marking. Do not tune to a test value; tune for performance.

On a street/strip-style engine, you test for the best fuel mixture, best timing, and to define the RPM ranges for torque and power.

A dyno provides a lot of data, assuming that all the sensors are operational and connected. Obviously, it provides horsepower and torque numbers, but it also provides fuel efficiency data as measured by the BSFC curve, and fuel mixture as defined by oxygen-sensor readings. These are both measuring fuel usage in different ways. You can tune with only one, but having both is a very good idea. There are no "most correct" numbers for either measurement, but there are commonly accepted ranges that you should expect to be within.

The BSFC measures the amount of fuel used to get 1 hp out of the engine. Typically it's between .350 and .600 at WOT. It used to be said that a BSFC number of .500 was a

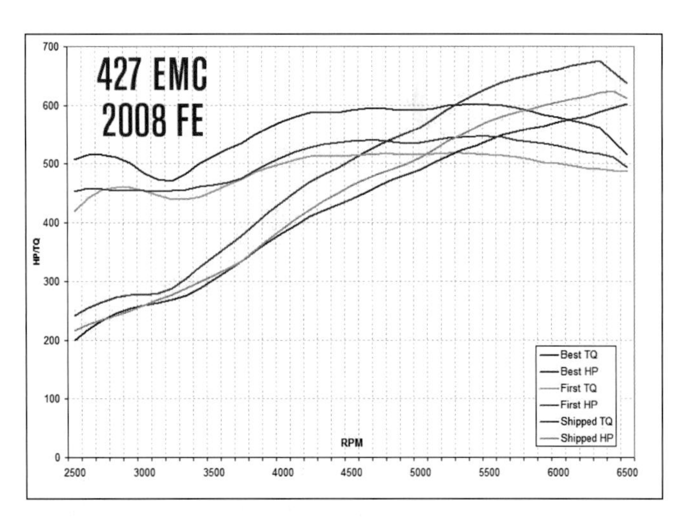

A great example showing the benefits of dyno tuning is this 2008 Engine Masters Challenge entry. This shows the first full pull, the results of tuning prior to shipping to the contest, and the final best output.

good one, but you commonly see far lower values in very good engines today. A lower value is considered leaner or more efficient as long as no detonation is present.

Wide-band oxygen sensors essentially measure the amount of oxygen in the exhaust stream. This correlates pretty well to the actual fuel mixture, and the data normally tracks along with the BSFC as the mixture is made leaner or richer. Output from the oxygen sensors is expressed as an "air/fuel ratio," with the ratio value going up as mixtures get leaner. Expectations of between 12.5:1 and 13.5:1 are reasonable at WOT for naturally aspirated engines.

Assuming that everything is operating properly, you are ready to make some real power—well, almost ready. Get the timing set to your baseline position, get the engine fully warmed up, and reset valve lash with the proper springs in place.

I'll start the engine and run it up to around 3,000 rpm to check total timing. I like to start out with a conservative value, somewhere around 32 degrees. I quickly scan the readouts for oil pressure, fuel pressure, and temperatures. If all is okay, I close the door to the dyno room and make the first real test pull. The first one is usually only to around 5,000 rpm but under full load. I am looking for a safe fuel curve and smooth torque delivery. If the data looks good, I make a few subsequent pulls in higher RPM increments until I establish the engine's RPM peak for horsepower. I like to see a clear roll-off for a couple hundred RPM at the top of the power curve; otherwise a small dip might fool you into thinking the engine is done when it has more yet to give.

Unless the fuel curve appears dangerous, I do a few timing loops first, finding the best power in 2-degree increments. Because most FE packages are comfortable somewhere between 28 and 38 degrees, a good timing baseline should be established within a half-dozen pulls.

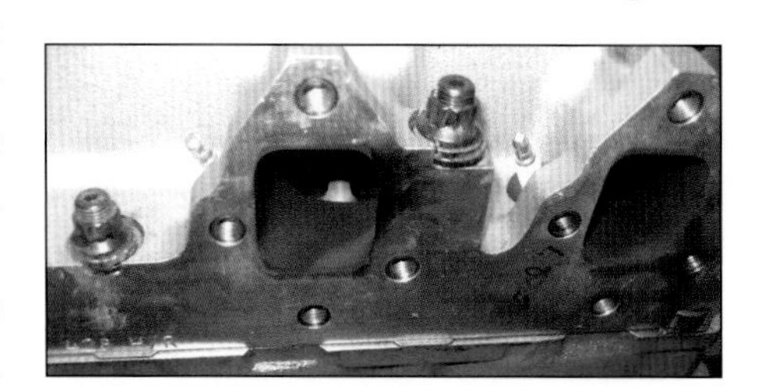

Exhaust-port color can be a good indicator of engine tune because the color tells you if the engine is rich, lean, or right on. This works with race gas, but unleaded 87-octane pump gas turns the ports sticky and black every time.

Getting control of the fuel curve can be more challenging. Most often, things come in pretty close to usable for a baseline on a new carburetor. I am not going to get into full-scale carburetor tuning because that's a book in itself, but I can provide a few working guidelines that will be helpful.

The total fuel-delivery curve is determined as a combination of flow through several circuits. The main circuit is responsible for the majority of fuel delivery at higher speeds, with an additional 20 to 40 percent of the total flow coming through the power-valve-channel restrictions. The idle circuit is still active even at WOT and delivers a fair percentage of the fuel package at lower RPM. However, its significance diminishes as the main circuit becomes fully active.

Changing the main jets alters the entire WOT fuel curve, as well as the steady state unloaded fuel delivery. Changing the size of the power

FE on the dyno from the rear, showing the hookup to the flywheel end. Visible is the black cooling tower in front, which is used to control engine water temperature and supply. The thick aluminum plate between the flywheel and the block is a rear mount for attaching the engine to the dyno cart. The flat plate bolted to the flywheel is part of the assembly coupling the engine to the dynomometer itself.

Here is a bad plug (left) and a good plug (right). You won't often see this with a new engine, but plugs are another good indicator of tune, and can serve to back up sensor data. A black or carboned plug is too rich; a white or light brown plug is too lean; and a milky brown color usually indicates correct jetting. But these are the most general of guidelines. Unleaded fuel won't give good visual data.

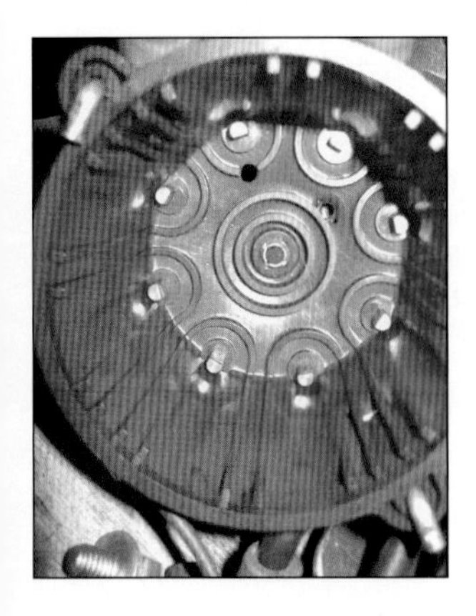

This extra hole in the distributor cap can reduce corrosion buildup on terminals and components.

valve-channel restriction alters the high-load and WOT parts of the fuel curve without adversely affecting fuel mixture at constant or steady state throttle application. Often it's better to keep the part-throttle lean and crisp by leaving the jetting slightly lean and opening up the power valve channel restrictions for peak power fuel. Changes to the main air bleeds affect both the timing of the main circuit (where it starts flowing) as well as the fuel-mixture quality and ratio, which bleeds "add air" into the fuel mix coming through the boosters.

Reading spark plugs is the "old school" way of monitoring fuel mixture, and still has some merit. The plug color can be impossible to read when running unleaded gas, though, because the fuel tends to color the shells black and leave the insulators bone white. I normally check the plugs after I think I've got the tune optimized as a cross-check for the data from the sensors. Remember: Tune for the best power, and use the sensors as a guide. You don't want to optimize the sensor readings; you want to optimize the engine!

With everything close to optimal, you can safely test the performance of a variety of parts, such as carb spacers, different headers, mufflers, air cleaners, plug gaps, and other timing settings, etc. But to eliminate variables, you should add one part or make one change at a time, so you are certain of the source of the change. The changes you make will be clearly reflected in the data, without the concern of damage or unknown variables.

Once satisfied, you can remove the engine from the dyno and get it ready for the car. Drain and inspect the break-in oil. It usually looks a bit dark and "sparkly" from break-in material, but it should otherwise be free of water or debris. Cut open the filter for final inspection. Check carefully for any leaks and fix them immediately. Mark your distributor for position in case it gets bumped during engine installation. And remember, you still need to tweak a bit once it's in the car to optimize the transient behavior.

Get out there and enjoy your new engine!

Most Holley parts (such as jets, gaskets, and power valves) are universal and, therefore, interchangeable to most Holley carb models. There is a lot more to professional tuning, but these simple parts usually get you pretty close when all parts are in good shape and you have good data.

A polished EFI-equipped 445 on the dyno shows the additional wiring and fuel connections required to run the fuel-injection system. This package is tuned with a laptop computer, and allows considerable control of all fuel- and ignition-related parameters.

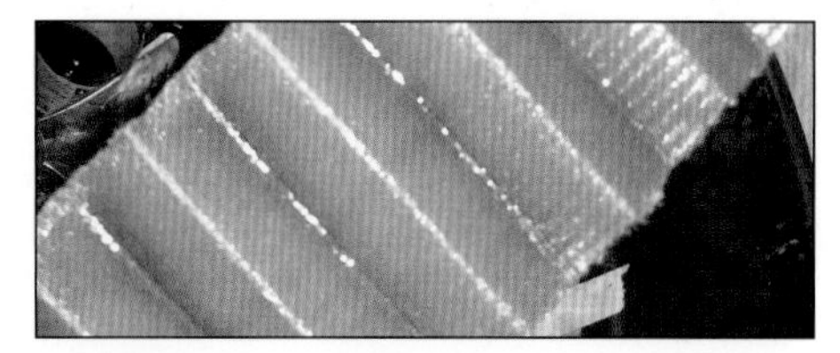

Oil filter inspection shows you any debris that accumulates during break-in. You normally see a small amount of debris; it should be only a couple of flecks of tiny stuff here and there, nothing large.

Valve deposits are an indication of oil intrusion. Bad valve seals or bad guides are the normal causes, but on an FE a poor intake gasket seal is also a common culprit.

SOURCE GUIDE

ARP
1863 Eastman Avenue
Ventura, CA 93003
800-826-3045
www.arpfasteners.com

ATI Performance Products
6747 Whitestone Road
Baltimore, MD 21207
877-298-5039
www.atiracing.com

Billet Specialties
500 Shawmut Avenue
La Grange, IL 60526
800-245-5382
www.billetspecialties

Blue Thunder
255 North El Cielo,
Suite 140-499
Palm Springs, CA 92262
760-328-9259
www.bluethunderauto.com

Canton Products
232 Branford Rd
Branford, CT 06471
203-481-9460
www.cantonracingproducts.com

Clevite Engine Parts
1350 Eisenhower Place
Ann Arbor, MI 48108
800-338-8786
www.clevite.com

Cloyes Gear & Products
7800 Ball Road
Fort Smith, AR 72908
479-646-4662
www.cloyes.com

Competition Cams
3406 Democrat Rd
Memphis, TN 38118
901-795-2400
www.compcams.com

Danny Bee Racing
30752 Imperial Street
Shafter, CA 93263
661-746-0517
www.tcwperformance.com/

Diamond Pistons
23003 Diamond Drive
Clinton Twp, MI 48305
877-552-2112
www.diamondracing.net

Doug Thorley
1180 Railroad Street
Corona, CA 92882
800-347-8664
www.dougthorleyheaders.com

Dove Manufacturing
27100 Royalton Road
Columbia Station, OH 44028
440-236-5139
www.doveengineparts.com

DSC Motorsports
59748 Reynolds Way
Anza, CA 92539
951-763-9765
www.dscmotorsport.com

Edelbrock
2700 California Street
Torrence, CA 90503
310-781-2222
www.edelbrock.com

Erson
16A Kit Kat Drive
Carson City, NV 89706
800-641-7920
www.pbmperformance.com

Federal-Mogul (Sealed Power,
Fel-Pro)
2555 Northwestern Highway
Southfield, MI 48075
248-354-7700
www.federal-mogul.com

FPA Headers
2526 23rd Avenue
Puyallup, WA 98371
253-848-9503
www.fordpowertrain.com

Genesis Castings
PO Box 19449
Indianapolis, IN 46219
317-357-8767
www.genesis427.com

Holley Performance
1801 Russellville Road
Bowling Green, KY 42101
270-781-9741
www.holley.com

Hooker
704 Highway 25 South
Aberdeen, MS 39730
270-781-9741
www.holley.com

Manley Performance Products
1960 Swarthmore Avenue
Lakewood, NJ 08701
732-905-3010
www.manleyperformance.com

Melling Oil Pumps
P.O. Box 1188
Jackson, MI 49204
517-787-8172
www.melling.com

Milodon
2250 Agate Court
Simi Valley, CA 93065
805-577-5950
www.milodon.com

Mondello Performance
Products
1103 Paso Robles Street
Paso Robles, California 93446
805-237-8808
www.mondellotwister.com

Moroso
80 Carter Drive
Guilford, CT 06437
203-458-0542
www.moroso.com

MSD Ignition
1350 Pullman Drive, Dock #14
El Paso, TX 79936
915-855-7123
www.msdignition.com

Precison Oil Pumps
2324 Decatur Avenue
Clovis, CA 93611
559-325-3553
www.precisionoilpumps.com

Probe Racing
2555 West 237th Street
Torrence, CA 90505
310-784-2977
www.probeindustries.com

Quick Fuel Technoogy
129 Dishman Lane
Bowling Green, KY 42101
270-793-0900
www.quickfueltechnology.com

Robert Pond Motorsports
1651 Coteau Drive
Riverside, CA 92504
909-376-2530
www.robertpondmotorsports.com

Scat Enterprises
1400 Kingsdale Avenue
Redondo Beach, CA 90278
310-370-5501
www.scatenterprises.com

Shelby Enterprises
19021 S. Figueroa Street
Gardena, CA 90248
310-538-2914
www.carrollshelbyenginecompany.com

Survival Motorsports
4202 Pioneer Drive, Suite E
Commerce, MI 48390
248-366-3309
www.survivalmotorsports.com

T&D Machine
4859 Convair Drive
Carson City, NV 89706
775-884-2292
www.tdmach.com

Total Seal
22642 North 15th Avenue
Phoenix, AZ 85027
623-587-7400
www.totalseal.com

Trend Performance
23444 Schoeherr
Warren, MI 48089
800-326-8368
www.trendperform.com

TWM Induction
325D Rutherford Street
Goleta, CA 93117
805-967-9478
www.twminduction.com